TABLE OF CONTENTS

Top 20 Test Taking Tips 4
Construction 5
    Geotechnical 33
    Structural 66
    Transportation 102
Water Resources and Environmental 124
Secret Key #1 - Time is Your Greatest Enemy 161
    Pace Yourself 161
Secret Key #2 - Guessing is not Guesswork 161
    Monkeys Take the Test 161
    $5 Challenge 162
Secret Key #3 - Practice Smarter, Not Harder 163
    Success Strategy 163
Secret Key #4 - Prepare, Don't Procrastinate 163
Secret Key #5 - Test Yourself 164
General Strategies 164
Special Report: How to Overcome Test Anxiety 170
    Lack of Preparation 170
    Physical Signals 171
    Nervousness 171
    Study Steps 173
    Helpful Techniques 174
Special Report: Additional Bonus Material 179

Copyright © Mometrix Media. You have been licensed one copy of this document for personal use only. Any other reproduction or redistribution is strictly prohibited. All rights reserved.

Top 20 Test Taking Tips

1. Carefully follow all the test registration procedures
2. Know the test directions, duration, topics, question types, how many questions
3. Setup a flexible study schedule at least 3-4 weeks before test day
4. Study during the time of day you are most alert, relaxed, and stress free
5. Maximize your learning style; visual learner use visual study aids, auditory learner use auditory study aids
6. Focus on your weakest knowledge base
7. Find a study partner to review with and help clarify questions
8. Practice, practice, practice
9. Get a good night's sleep; don't try to cram the night before the test
10. Eat a well balanced meal
11. Know the exact physical location of the testing site; drive the route to the site prior to test day
12. Bring a set of ear plugs; the testing center could be noisy
13. Wear comfortable, loose fitting, layered clothing to the testing center; prepare for it to be either cold or hot during the test
14. Bring at least 2 current forms of ID to the testing center
15. Arrive to the test early; be prepared to wait and be patient
16. Eliminate the obviously wrong answer choices, then guess the first remaining choice
17. Pace yourself; don't rush, but keep working and move on if you get stuck
18. Maintain a positive attitude even if the test is going poorly
19. Keep your first answer unless you are positive it is wrong
20. Check your work, don't make a careless mistake

Construction

Excavation

Excavation is defined as any man-made removal of earth materials. The four major classifications of soil listed in decreasing order of stability are Stable Rock, Type A, Type B, and Type C.
- Type A soils have unconfined, compressive strengths of 1.5 tons per square foot (tsf) or greater. Type A soils are cohesive soils; examples include clay, silty clay, and sandy clay.
- Type B soils have unconfined, compressive strengths of greater than 0.5 tsf to less than 1.5 tsf. Type B soils include both cohesive and cohesion-less soils. Examples of Type B soils include angular gravel, silts, sandy loams, and medium clays.
- Type C soils are the most unstable soils and have unconfined, compressive strengths of 0.5 tsf or less. Examples of Type C soils include gravel, sand, loamy sand, submerged soil, and submerged unstable rock.

Sloping and benching

Sloping and benching is a method of protecting an excavated area from caving in or collapsing. It may be used in lieu of shoring to protect the work area and the employees working within an excavation. Stable Rock has a maximum allowable slope of 90 degrees (vertical). Type A soils have a maximum allowable slope of 53 degrees from the horizontal, or a horizontal to vertical ratio of 3/4:1. Type B soils have a maximum allowable slope of 45 degrees from the horizontal, or a horizontal to vertical ratio of 1:1. Type C soils have a maximum allowable slope of 34 degrees from the horizontal, or a horizontal to vertical ratio of 1-1/2:1. These slopes cannot be used at excavation depths greater than 20 feet; for depths greater than 20 feet, the sloped walls must be designed by a professional engineer. No sloping is generally required for excavations less than 5 feet deep.

Soil distress

Soil distress occurs when soils exhibit signs of potential collapse or cave-in. Several causes of soil distress include freezing or subsequent thawing of soil, water seepage from rain or some other water source, high temperatures combined with low humidity, and disturbance of soil caused by nearby vibrating machinery or moving equipment. Indicators of soil distress include loosening and trickling of material at walls of excavation, fissures or cracks in soil, sinking of material at edge of excavation, material bulging or heaving at base of excavation, and material slumping at the face of the excavation walls. The Occupational Safety & Health Administration (OSHA) federal agency mandates that the maximum allowable slopes be reduced by 50% if the soil exhibits signs of distress. OSHA also mandates that spoils and equipment be kept at least 2 feet from the edge of excavation to help prevent soil distress (29 CFR 1926).

Cut and fill

Determination of cut and fill volumes making up the embankment is essential in almost every construction site development project. A cut refers to the amount of material removed at a particular location on the site. An embankment or fill is a raised area that supports a nearby feature, such as a roadway or railway, and refers to the amount of material needed to bring the structures or improvements up to the proposed grade. Ideally, the cut and fill volumes would be near identical to limit costs associated with hauling and dumping the extra material and costs associated with purchasing and bringing additional material onto the site. The slopes of a cut section generally are not steeper than 27 degrees from the horizontal, or a horizontal to vertical ratio of 2:1. For areas that are filled, it is generally recommended that the maximum fill height not exceed 20 feet.

Mass diagram

A mass diagram is a diagram of the earthwork volume plotted cumulatively along an alignment. A mass diagram helps determine the cut and fill volumes and can help determine a finished elevation that balances the two. The station distance is plotted along the x-axis. The cumulative quantity of cut and fill volumes is plotted along the y-axis. A line with a positive slope on the mass diagram indicates an area of cut (excavation), and a line with a negative slope on a mass diagram indicates an area of fill (embankment). A straight, horizontal line between two points on a sag or crest in the mass diagram is called a balance line. A balance line signifies that the excavation and embankment volumes between the two endpoints of the balance line are equal. Soil shrinkage and expansion for fill and cut material can be accounted for in a mass diagram by increasing the respective fill or cut quantity by a predetermined percentage.

Problem

Calculate the volume (in cubic yards) of the excavation shown below using the borrow pit method (also known as the grid method). The proposed grade is 150.0 feet. The area of each grid square is 2025 square feet.

Solution

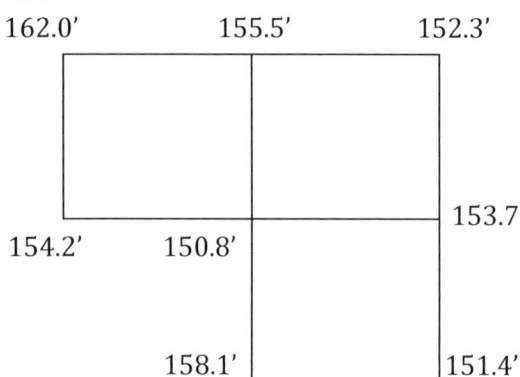

The formula for the borrow pit method (grid method) is:

$$V = \sum (h_{i,j} n) \left(\frac{A}{4 \times 27} \right)$$

Where V is the volume of the excavation in cubic yards, $h_{i,j}$ is the height in feet above the desired grade at row i and column j, n is the number of corners at point i,j, and A is the area of each grid in square feet.

Solution: The height in feet above the proposed grade, $h_{i,j}$, at each grid square corner is shown below.

$$V = (12.0 \times 1 + 5.5 \times 2 + 2.3 \times 1 + 3.7 \times 2 + 1.4 \times 1 + 8.1 \times 1 + 0.8 \times 3 + 4.2 \times 1)\left(\frac{2025}{4 \times 27}\right)$$

$$V = 915.0 \text{ cy}$$

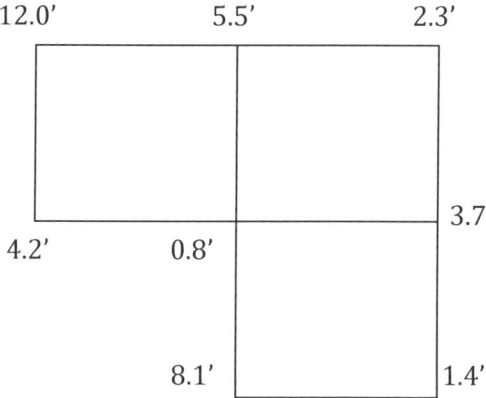

Problem

Use the average end area method to calculate the earthwork volume for the example below. Discuss limitations of this method.

Solution

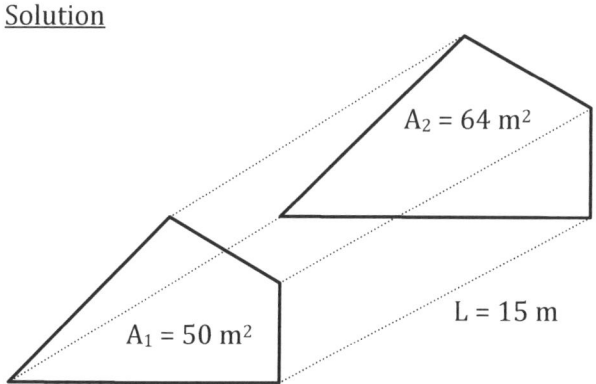

The average end area method for calculating earthwork volume is determined by averaging the two end areas and multiplying by the soil prism length:

$$V = \frac{(A_1 + A_2)}{2} * L = \frac{(50m^2 + 64m^2)}{2} * 15m = \mathbf{855 m^3}$$

The average end area method is conservative and in general sufficient for most earthwork calculations. This method should not be used when a) the end area(s) is/are complex, and/or b) one end of the soil prism is extremely small or near zero. For complex end areas, the end areas may be plotted on graph paper and the squares counted to determine the areas. For truncated soil prisms, the earthwork volume can be determined by using the formula for a pyramid:

$$V_{pyramid} = \frac{A_{base}}{3} * L$$

Importance of a construction site layout

A suitable construction site layout is critical to the success of a project. A poorly planned construction site layout can lengthen a project schedule, cause unnecessary disruptions, and increase project costs due to loss of productivity. Project, time, and construction trade constraints need to be taken into consideration to allow for a smoother transition among the phases of construction. An example of this would be designing adequate space for temporary electrical systems to be constructed and operated while the permanent electrical systems are being constructed, and meeting the planned transfer date as agreed to by the owner and electrical company. Access roads within the site layout boundary need to be adequate for moving large equipment to, from, and within the site without greatly restricting vehicular and pedestrian traffic. A good site layout will have several access roads available to the construction areas to limit interruptions due to common construction activities such as installation of underground utilities. A site layout must also be properly graded to avoid flooding of work areas.

Construction jobsite safety

The federal Occupational Safety and Health Administration (OSHA) requires that all chemicals and their hazards on a site be known to employees. A Material Safety Data Sheet (MSDS) must be readily available and accessible for each chemical. An MSDS contains the manufacturer's emergency contact information, identifies the chemical's hazardous ingredients and its chemical and physical characteristics, describes the reactivity and stability of the chemical, and also provides safe handling precautions. All employees working in confined spaces, such as manholes, vaults, and tanks, must be properly trained. The atmosphere within the confined space must be monitored to ensure adequate oxygen is available. Should oxygen levels not be sufficient, a breathing apparatus must be provided and worn by the employee. Fall protection is also required by OSHA for protection of employees on a jobsite. Fall protection has many forms. The most common forms are barricades at excavation boundaries, anchored personal fall-arrest systems, and head protection (hard hats). Exposed rebar ends must also be properly covered to prevent impalement during a fall.

Surveying terms

A *backsight (BS)* is a survey leveling rod reading taken from a point back to a benchmark or turning point with a known elevation. A *foresight (FS)* is a reading taken with the instrument at the same location as when taking the backsight reading but now pointed forward to a turning point with an unknown elevation. A *benchmark (BM)* is a point at which the elevation is known. The *height of instrument (HI)* is the elevation at which the instrument is located with respect to the benchmark and turning point elevations. A *turning point (TP)* is a point with an unknown elevation. Elevations of turning points are calculated.

The backsight leveling rod reading is added to the known elevation of a benchmark or turning point to determine the height of instrument: BS + BM or TP Elevation = HI
The foresight leveling rod reading is subtracted from the height of instrument to determine the elevation of the turning point: HI - FS = TP Elevation

Problem

Calculate the missing information for A, B, C, and D in the survey field log below:

Solution

Station	BS	HI	FS	Elevation
BM1	A	247.71		245.62
TP1	3.12	C	10.38	B
TP2	2.34	232.83	D	230.49
BM2			4.65	228.18

For A: to calculate the backsight (BS) at *BM1*, use the following equation:

BS + *BM1* Elevation = HI. Solving for BS at *BM1*: A = HI − *BM1* Elevation.

A = 247.71 − 245.62; **A = 2.09**

For B: to calculate the elevation at *TP1*, use the following equation:

TP1 Elevation = HI − FS; B = 247.71 − 10.38; **B = 237.33**

For C: HI = BS + *TP1* Elevation: C = 3.12 + 237.33; **C = 240.45**

For D: FS = HI − *TP2* Elevation: D = 240.45 − 230.49; **D = 9.96**

To check the answers, subtract the summation of the foresights from the summation of the backsights, and compare to the change in elevation from *BM1* to *BM2*. The numbers should be the same.

Check: (2.09 + 3.12 + 2.34) - (10.38 + 9.96 + 4.65) = -17.44, or a decrease in elevation of 17.44 feet.

245.62 − 228.18 = 17.44, so the numbers calculated are correct.

Construction stake information

The front of a construction stake is typically marked with what is known as header and cluster information. Header information located at the top of the stake identifies the offset with respect to the natural ground surface at the point the stake is driven and the object the offset is measured from (power pole, sidewalk, etc). Header information is typically separated from cluster information by a heavy line(s) or colored flagging. The cluster information identifies horizontal and vertical measurements. Multiple clusters may be marked on one stake, but all clusters are measured from the same point and in the same direction as the other clusters on the stake. The side of the construction stake facing away from construction (stake back) is marked with the station and any other descriptions. The width face or edge of the stake is marked with actual elevations if known. Information on a stake should be read from the top down. The location of the markings on a stake is ultimately dependent on the project standards.

Types of stakes
A *ground stake* is a stake without any markings that is driven flush with the existing grade. Sometimes a guard stake will be driven in at an angle over the ground stake to help protect and identify the location of the ground stake. A *witness stake* is a stake that references the location of a point with respect to a ground stake, located nearby. The witness stake does not contain any information related to the location of the witness stake itself. A *slope stake* indicates changes to be made in the elevation with respect to the existing grade (cuts or fills). Slope stakes are typically marked with an 'SS,' and cuts and fills will be identified by a 'C' or an 'F,' respectively. An *offset stake* is driven outside of the edge of an excavation (at an offset) so that during the excavation, the stake will not be disturbed. The offset distance is located within the header information on a stake and is typically circled.

Construction staking abbreviations
CB is the abbreviation for a catch basin, an inlet into a storm drainage system. The inlet is installed at a low point in a paved area or along a curb. CL refers to a centerline. Stations are typically set up along the centerline of a roadway. CR is the abbreviation for a curb return, the curved section of a curb where two streets intersect. EC is the abbreviation for end of curve. An end of curve is the point at which a curve ties into a straight line or into another curve of a different radius. EOP is edge of pavement. An example of this would be the paving edge of a roadway. INV refers to the invert of a structure or pipeline. The term invert is synonymous with the term flow line when referring to pipelines. SS is the abbreviation for either slope stake or sanitary sewer, depending on the project. A slope stake indicates a change in grading and sanitary sewer refers to a pipeline conveying sewage.

Features located on a topographic survey map

A topographic survey map is a map that indicates the elevation changes within an area using contour lines. Contour lines indicate a specific elevation (i.e., 100 feet), and usually parallel each other. Contour lines are shown at consistent elevation intervals, depending on the area's terrain. Contour lines that are close together indicate an area with a steep slope, and contour lines spaced further apart indicate areas with a milder slope. Hills and depressions will commonly appear the same on a topographic map, but can be differentiated by the elevation shown on a contour line or a spot elevation shown at the top or bottom of the hill or depression. In creating a topographic survey, the site is divided into equally-sized grids. The elevations at the four corners of the entire area are measured, and the elevations at each individual grid corner are determined based on the edge slopes of the area. The contour lines can then be created by using linear interpolation for the elevations within each grid.

Problem

Draw a profile of the proposed waterline elevation with respect to the witness stake shown.

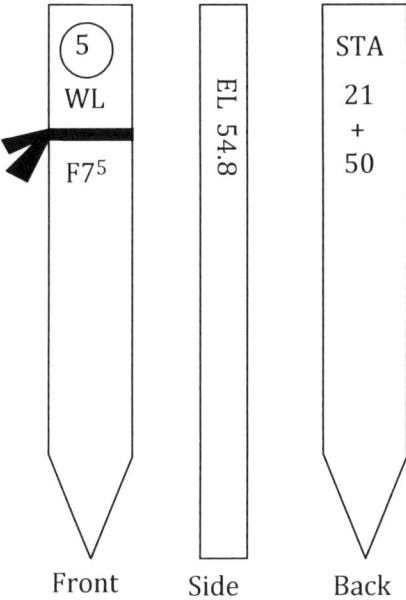

Front   Side   Back

Solution

The header information on the front of the witness stake indicates that the centerline of the proposed waterline will be at a 5 foot offset from the ground stake. The cluster information indicates that the invert of the water line will be 7.5 feet above the ground stake. Since the elevation of the ground stake is shown on the side of the stake, the elevation of the waterline invert can be calculated. The stationing shown on the back of the stake gives the station along the pipeline at this particular point.

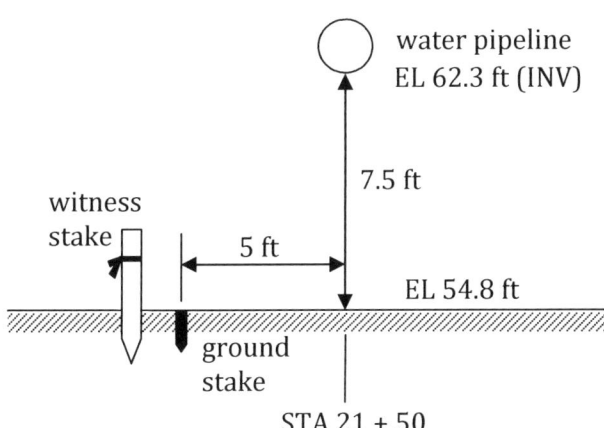

Problem

Sketch a profile for the proposed railway finished grade based on the information shown on the witness stake. A leveling instrument pointed at a ground rod placed at the ground stake has a reading of 4.5 feet. Determine the reading of a rod placed at the finished grade of the railway as taken by the same leveling instrument.

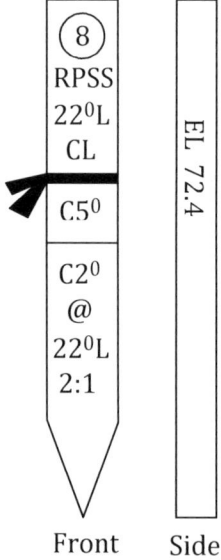

Solution

The profile for the finished grade of the proposed railway is shown below. The reading of the finished grade rod can be determined by adding the ground rod reading (4.5 feet) to the distance from the ground to the finished grade (7 feet). The finished grade rod reading would be **11.5 feet**.

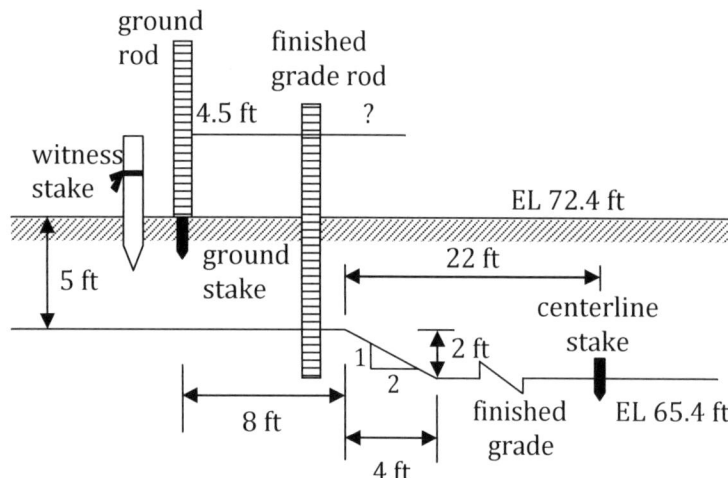

Problem

In measuring a property edge, the mean of ten separate measurements was 1,025 feet. Determine the upper and lower 50%, 95%, and 99% confidence limits for these measurements, given that the standard deviation for this set of numbers was 1.1 feet. Assume data to be normally distributed.

Solution
The equation for determining the standard error of the mean is given below:

$$SE_{mean} = \frac{(Q_{ND})(s)}{\sqrt{n}}$$

Where SE_{mean} is the standard error of the mean, Q_{ND} is the normal distribution quantile, s is the sample standard deviation, and n is the number of measurements. The normal distribution quantile for 50%, 95%, and 99% confidence limits are 0.6745, 1.96, and 2.57, respectively. For 50% upper and lower confidence limits, the equation becomes:

$SE_{mean} = \frac{(0.6745)(1.1)}{\sqrt{10}} = 0.23$ feet, indicating that the upper limit is 1,025 ft + 0.23 ft = 1025.23 ft, and the lower limit is 1025 − 0.23 feet = 1024.77 ft.

Solving similarly, the 95% upper and lower confidence limits are 1025.68 ft and 1024.32 ft, respectively. The 90% upper and lower confidence limits are 1025.89 ft and 1024.11 ft, respectively.

Problem

A surveyor's field notes lists the following interior angle measurements, taken just once, for a quadrilateral area: Angle A: 104°, Angle B: 80°, Angle C: 97°, and Angle D: 76°. Calculate the most probable interior angles for this quadrilateral.

Solution
The interior angles of any quadrilateral should add up to 360°. The error in these measurements is: 360° − (104°+80°+97°+76°) = 3°. The error is distributed equally to the four measurements to determine the most probably values of the interior angles as shown below.

For Angle A: $104° + \frac{3°}{4} = 104.75°$
For Angle B: $80° + \frac{3°}{4} = 80.75°$
For Angle C: $97° + \frac{3°}{4} = 97.75°$
For Angle D: $76° + \frac{3°}{4} = 76.75°$
Check: 104.75° + 80.75° + 97.75° + 76.75° = 360° ✓

Problem

An 880-foot neighborhood park is measured with a 155-ft tape with a probable error of 0.17 feet, but then re-measured with a 130-ft tape with a probable error of 0.15 feet. Calculate the expected error of each.

Solution
The total expected error for all of the individual measurements can be determined by the following equation:

$$E_{total} = SE_{mean} \times \sqrt{n}$$

Where E_{total} is the total error for all the measurements, SE_{mean} is the standard error of the mean, and n is the number of measurements. The number of measurements for this example can be determined by dividing the total length measured by the length of the tape. For the 155-ft tape, n = 880 ft / 155 ft = 5.68 or 6 measurements. For the 130-ft tape, n = 880 ft / 130 ft = 6.77 or 7 measurements. The equation becomes:

$$\text{For the } 155ft \text{ tape: } E_{total} = 0.17\ ft \times \sqrt{6} = 0.42\ ft$$

$$\text{For the } 130ft \text{ tape: } E_{total} = 0.15\ ft \times \sqrt{7} = 0.40\ ft$$

This indicates that using the 130-ft tape would be slightly more accurate (by 0.02 ft).

Problem

The distance between an engineering level and a leveling rod is 990 feet, accounting for the curvature of the earth. Determine the corrections to the observed rod reading due to earth curvature and atmospheric refraction.

Solution
A rod reading will appear higher than the actual reading due to the curvature of the earth. The correction due to the curvature of the earth is determined by the following equation:

$$C_{earth} = \left(2.4 \times 10^{-8} \frac{1}{ft}\right) x^2$$

where x is the distance in feet from the level to the rod, accounting for earth curvature. The equation becomes:

$$C_{earth} = \left(2.4 \times 10^{-8} \frac{1}{ft}\right)(900ft)^2 = 0.019\ ft\ or\ 0.02\ ft$$

A rod reading will appear lower than the actual reading due to atmospheric refraction. The correction due to atmospheric refraction is determined by the following equation:

$$C_{refraction} = \left(3.0 \times 10^{-9} \frac{1}{ft}\right) x^2 = \left(3.0 \times 10^{-9} \frac{1}{ft}\right)(900ft)^2 = 0.002\ ft\ or\ 0\ ft$$

The actual reading of the rod is determined by adding $C_{refraction}$ and subtracting C_{earth} to the observed rod reading.

Problem

The interior angles of a quadrilateral area were measured several times. Angle A's measurement of 37° was taken 4 times, Angle B's measurement of 125° was taken 5 times, Angle C's measurement of 46° was taken 3 times, and Angle D's measurement of 154° was taken 2 times. Determine the probable value for each angle.

Solution
The sum of a quadrilateral's interior angles should add up to 360°. The sum of the measured angles is 362°, a difference of -2°. To correct the interior angles, the first step is to add up the inverses of each measurement: $\frac{1}{4} + \frac{1}{5} + \frac{1}{3} + \frac{1}{2} = 1.283$

This correction is applied to each interior angle by the following equation:

$$\angle_{interior} + \frac{\left(\frac{1}{n}\right)(C)}{\sum inverse} = \angle_{corrected}$$

Where *n* is the number of measurements for the interior angle being corrected, *C* is the total interior angle correction difference, and $\sum inverse$ is the sum of all the interior angle measurement inverses.

$$\angle A_{corrected} = 37° + \frac{\left(\frac{1}{4}\right)(-2°)}{1.283} = 36.61°$$

Solving similarly for all the remaining interior angles, the corrected interior angles are:

$\angle B_{corrected}$ = 124.69°, $\angle C_{corrected}$ = 45.48°, $\angle D_{corrected}$ = 153.22°.

Global Positioning System (GPS)

GPS is a navigation system that utilizes satellite-based equipment to determine the position of a point on earth. A receiver is held on a particular spot on earth and receives information from satellites that is used to calculate the receiver's position. At least four satellites must be clearly visible to the receiver to produce an accurate position. A main advantage of using GPS for land surveying includes the relative speed and ease of position acquisition with less chance of human error compared to other more traditional, labor-intensive methods. GPS also has the ability to be used during nighttime or periods of poor weather, as long as four satellites are visible. Disadvantages of GPS include its relative high cost and its reliance on four available satellites. GPS is not the best choice for measuring elevations due to the difference between the actual shape of the earth and the mathematical model used in the system's software.

Problem

Sketch the location of a parcel of land with the following partial legal description: the NW ¼ of the SW ¼ of the NE ¼ of the NE ¼ of Section 15.

Solution
Section 15 of a township is shown below. Start backwards with the NE ¼ identifier. This indicates that the parcel of land is in the northeast quadrant of Section 15. Continue with the identifiers to the left until the correct parcel of land is identified, as indicated below.

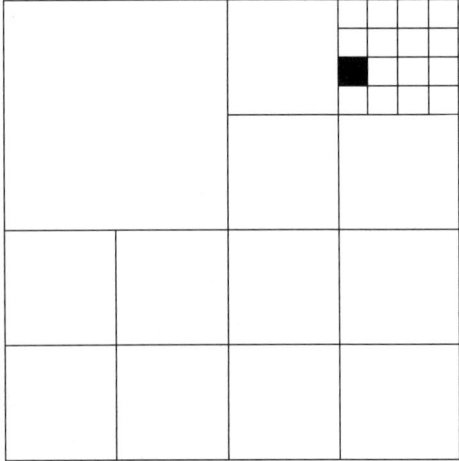

Problem

Determine the film size of aerial mapping equipment if a photograph covers an area of 100,000,000 ft² and a scale of 1 inch to 2,000 feet is being used. Determine the sidelap and endlap if the distance between flight paths is 6,000 ft, and the distance between photos along the flight path is 4,100 ft.

Solution
Use the following equation to solve for the film size:

$$100,000,000 ft^2 = \left(\frac{2000\ ft}{1\ in}\right)^2 \times A_{photo}\ ;\ A_{photo} = 25 in^2$$

The size of the film is thus 5 inches x 5 inches.

The sidelap is the predetermined overlap in flight lines adjacent to one another. An endlap is the predetermined overlap between the photos along the same flight line. Each photograph taken covers an area of 100,000,000 ft², or 10,000 ft by 10,000 ft.

The sidelap can be determined from the following equation for this problem:

$$(1 - sidelap)(10,000 ft) = 6,000\ ft\ ;\ sidelap = 0.40\ or\ 40\%$$

Similarly, the endlap can be determined using the equation below:

$$(1 - endlap)(10{,}000\,ft) = 4{,}100\,ft\,;\,endlap = 0.59\,or\,59\%$$

Problem

Discuss the corrections taken into account when using a surveyor's tape for horizontal distances and give equations for each.

Solution
Corrections due to tape sagging, thermal expansion, incorrect tape length, and tension all need to be taken into account when determining the true length of a horizontal distance measured with a tape. The correction equations are shown below:

$$C_{sag} = \frac{W^2 L_f^3}{24 P^2}\,;\,C_{temp} = L_f \propto (T - T_s)$$

$$C_{Length} = L_f \left(\frac{L_t - L_s}{L_s}\right)\,;\,C_{tension} = L_f \left(\frac{P - P_s}{AE}\right)$$

Where W is the weight of the tape per unit length, L_f is the measured distance obtained in the field, P is the force applied to the tape during measurement, α is the thermal expansion coefficient for the tape material, T is the temperature at which the measurement was taken, T_s is the temperature during standardization, L_t is the tested tape length (as compared to L_s), L_s is the length of the tape during standardization, P_s is the force applied to the tape during standardization, A is the cross-sectional area of the tape, and E is the tape's modulus of elasticity.

Azimuths

Azimuths are angles measured typically from the north end of a meridian in a clockwise direction, but they may also be measured from the south end of a meridian. Azimuths can never exceed 360°. Bearings are angles measured from the north or south end of the meridian and referenced by the quadrant in which they are located. Bearings are identified beginning with an N or S (for North or South) and ending with a W or E (for West or East). Bearings cannot exceed 90°.

Problem

Determine the bearings for the azimuths shown below.

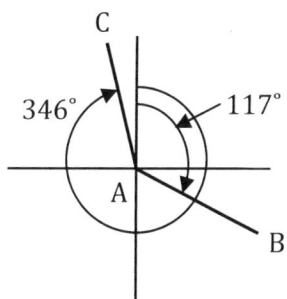

Solution

The bearing for line AB will be measured from the south end of the meridian (since bearings cannot exceed 90°). The bearing angle can be calculated by subtracting its azimuth from 180°: 180° - 117° = 63°. The bearing for line AB is thus S 63° E.

The bearing angle for line AC is measured from the north end of the meridian and calculated by subtracting its azimuth from 360°: 360° - 346° = 14°. The bearing for line AC is thus N 14° W.

Problem

Two specific benchmarks are shown on three separate maps, all with different scales. The scale of the first map is 1 inch: 1,000 feet. The benchmarks on this map are shown 2.5 inches apart. The second map has a scale of 1 inch: 300 feet and the third map has a scale of 1 inch: 2,800 feet. Determine how far apart the benchmarks should be shown from each other on the second and third maps.

Solution

The scales and distances between objects on two maps are related to each other by the following equation:

$$\frac{Scale_{MapA}}{Scale_{MapB}} = \frac{Distance\ between\ objects_{Map\ B}}{Distance\ between\ objects_{MapA}}$$

For the second map, the equation becomes:

$$\frac{\frac{1000 ft}{1\ in}}{\frac{300\ ft}{1\ in}} = \frac{Distance\ between\ objects_{Map\ B}}{2.5\ inches}$$

The distance between the benchmarks on the second map (with a scale of 1 inch: 300 ft) is 8.33 inches.

For the third map, the equation is:

$$\frac{\frac{1000 ft}{1\ in}}{\frac{2800\ ft}{1\ in}} = \frac{Distance\ between\ objects_{Map\ C}}{2.5\ inches}$$

The distance between the benchmarks on the third map (with a scale of 1 inch: 2,800 ft) is 0.89 inch.

Problem

Determine the height of a tower in an aerial mapping photograph if the tower displacement measured is 0.24 inches, the radial distance from the nadir to the top of the tower on the photograph is 3.0 inches, the scale of the photograph is 1 inch: 550 feet, and the focal length of the camera is 5.5 inches.

Solution
The first step in solving this problem is to calculate the altitude of the flight above the datum using the following equation:

$$scale = \frac{focal\ length}{A_{flight}}; A_{flight} = \frac{focal\ length}{scale}$$

$$A_{flight} = \frac{5.5\ inches}{\frac{1\ inch}{550\ ft}} = 3025\ ft\ \text{above the base of the tower}$$

The altitude is then used in the following equation to solve for the height of the tower, where D is displacement of the object in inches as measured in the photograph and r is the radial distance from the nadir to the top of the displaced object:

$$H_{object} = \frac{D * A_{flight}}{r} = \frac{(0.24\ in)(3025\ ft)}{3.0\ in} = 242\ ft$$

The height of the tower is near 242 feet.

Problem

Line AB has a latitude of -124.82 ft and a departure of +341.74 ft. Line BC has a latitude of +259.28 ft and a departure of +204.33 ft. Determine the value of interior Angle B.

Solution
A line's latitude runs north and south. A north-running latitude is positive, and a south-running latitude is negative. A line's departure runs west and east. A west-running departure is negative, and an east-running departure is positive. Lines AB and BC are shown below with their corresponding latitudes and departures. The first step is to determine θ1 and θ2.

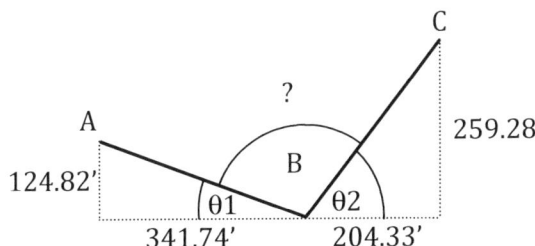

tan θ1 = (124.82/341.74); θ1 = 20.06°

tan θ2 = (259.28/204.33); θ1 = 51.76°

The interior angle at B thus equals 180° - 20.06° - 51.76° = 108.18°

Methods for estimating project quantities.

The *unit quantity take-off method* is a detailed method that involves separating the project into individual tasks and determining the quantity of materials, equipment, tools, labor, etc. needed to complete each task using a specific unit of measurement. Quantities are determined using the project drawings and specifications. The cost per unit of measurement is determined and then multiplied by the quantity to obtain a cost for that specific item. The costs for all items are added together to obtain a total project cost estimate. A benefit of this method is that costs per unit of measurement may easily be obtained from previous projects and changed to reflect current conditions.

The *total quantity take-off method* is a detailed method that involves separating the project into five groups: Materials, Labor, Plant, Overhead, and Profit. The quantity within each group is determined through review of the project drawings and specifications and multiplied by the group's unit cost. The costs of each group are added together for the total project cost estimate.

Methods for estimating project costs

The "*bottoms up*" method relies on project drawings and specifications for quantity estimation. This method involves careful review of the drawings and specifications to determine quantities of material, tools, and equipment needed for each task within the project. This method may also incorporate the aid of engineering design software to determine quantities. Once quantities are attained, costs associated with materials, equipment, labor and overhead can then be determined. Successful use of this method is highly dependent on the level of detail and thoroughness of the project drawings and specifications.

The *analogy take-off method* relies on known costs of similar past projects or systems. Costs may then be adjusted based on specific project differences or circumstances. This method is highly useful for cross-checking other cost estimating methods or when detailed project information is not available.

The *parametric cost estimating method* relies on historical data of similar projects or systems and/or developed models in which, through statistical analysis, cost relationships between parameters can be determined and established, then subsequently applied to determine costs for the current project. This method may be used when detailed project information is unavailable. The *trends analysis cost estimating method* relies on an index created by comparing total project costs to that of work currently completed. This gives the project manager and contractor a good idea of whether the project is on budget. Costs of subsequent work can then be changed if necessary. When detailed information is lacking for a project, the *expert opinion method* for estimating project costs can be employed. Expert individuals with vast experience and knowledge in similar projects are hired to review the project's design and collectively determine a cost estimate for the project.

Problem

A contractor purchases a generator for $2,100. The annual operations and maintenance cost for the generator is $250. The contractor expects to make $600 each year of the generator's life. *Draw the cash flow diagram and determine at what point the contractor would break even.*

Solution
The cash flow diagram is shown below and could be simplified further by combining the expenses and revenues for each year (net revenue of $350 each year).

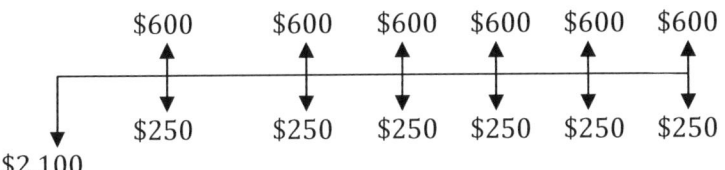

The contractor would break even at the end of year 6. This can be calculated by the following simple equation:

$$\frac{\$2{,}100}{(\$600 - \$250)} = 6, or\ at\ the\ end\ of\ the\ sixth\ year$$

Nominal interest rate, effective interest rate, and single payment compound amount factor

A nominal interest rate refers to an interest rate that does not take into account inflation. Nominal interest rates are not generally used to calculate equivalent values as the compounding periods do not coincide with the interest rate period. An effective interest rate refers to an interest rate that may be used in calculations as the compounding periods do coincide with the interest rate period. The single payment compound amount factor is a factor used to calculate future worth from present worth. The equation for the factor is (1+i)n and is typically pre-calculated as a tabulated value.

The future worth can be calculated from the following equation:

$$F = P(1 + i)^n$$

Where *F* is the future worth, *P* is the present worth, *i* is the effective interest rate, and *n* is the number of compounding periods.

Problem

Calculate the future worth of a $35,000 contribution 9 years from now, assuming an effective annual interest rate of 6%.

Solution

$$F = P(1 + i)^n = \$35{,}000(1 + 0.06)^9 = \$59{,}132$$

Problem

A designer is evaluating two advanced water treatment pilot systems for ongoing research. Option A has the following characteristics: initial cost of $4,000, a life of 6 years, and an annual maintenance cost of $35. Option B has the following characteristics: an initial cost of $1,500, a life of 2 years, and an annual maintenance cost of $65. Determine which option is better assuming an interest rate of 5%.

Solution
In evaluating options with unequal lives, it is necessary to use the annual cost method to determine which option is best.

The equations for calculating the equivalent annual cost (EAC) are shown below for each option:

$$EAC = AC + P\left(\frac{i(1+i)^n}{(1+i)^n - 1}\right)$$

Where AC is any annual cost, P is the present worth, i is the effective interest rate, and n is number of compounding periods.

$$EAC_{Option\ A} = \$35 + \$40,000\left(\frac{0.05(1+0.05)^6}{(1+0.05)^6 - 1}\right) = \$823$$

$$EAC_{Option\ B} = \$65 + \$1,500\left(\frac{0.05(1+0.05)^2}{(1+0.05)^2 - 1}\right) = \$872$$

Option A is the better option of the two because of its lower annual cost.

Problem

A steel worker is hired by a contractor to work five days a week, twelve hours a day at a rate of $30 per hour. The steel worker has additional costs that include:
  a) FICA (6.2%),
  b) unemployment insurance of 4%,
  c) public liability and property damage insurance of $4.50 per $100 of pay,
  d) worker's compensation of $4 per $100 of pay, and
  e) fringe benefits of $1 per hour. Determine the hourly labor cost that the contractor pays for this steel worker. Assume no overtime/shift differential or subsistence pay.

Solution
Base pay: the steel worker works 60 hours each week (12 hours per day, 5 days a week). His weekly base pay is $30/hour x 60 hours/week = $1800.

Additional compensation:
  a) FICA: 0.062 x $1800 = $111.60 per week
  b) Unemployment insurance: 0.04 x $1800 = $72 per week
  c) Public liability and property damage insurance and workers compensation: $4.50 per $100 of pay: $4.50 x ($1800/$100) = $81 per week

d) Worker's compensation: $4 per $100 of pay: $4 x ($1800/$100) = $72 per week
e) Fringe benefits: $1/hour x 60 hours/week = $60 per week

The total pay per week is thus $1800 + $111.60 + $72 + $81 + $72 + 60 = $2196.60, and the total hourly labor cost for the steel worker is $2196.60/60 hours = $36.61/hour.

Problem

A conical spoil pile has a diameter of 40 feet. The material has an angle of repose of 32° and a 10% swell. Determine 1) the volume of the natural material prior to its deposit as spoil, and 2) the number of dump truck loads that would be required to remove the spoil pile assuming the dump truck can haul 20 loose cubic yards of the material at a time.

Solution
The volume of the conical spoil pile can be calculated from the formula for a cone:

$$Diameter_{pile} = \left(\frac{24 \times Volume_{pile}}{\pi \tan R}\right)^{\frac{1}{3}}$$

Where R is the angle of repose.

$$40 \, feet = \left(\frac{24 \times Volume_{pile}}{\pi \tan R}\right)^{\frac{1}{3}}$$

$$Volume_{pile} = 5{,}235 \, cubic \, feet \, (cf)$$

1) The volume of the natural material prior to swelling:

5,235 cf ÷ 1.10 = 4,759 cf (or 176 cy).

2) The number of dump truck loads would be based on the material with swelling:

Convert cubic feet to yards: 5,235 cf ÷ 27 cf/cy = 194 cy of material with swelling

Calculate the number of loads: 194 cy ÷ 20 cy = 9.7 loads

Thus, 10 dump truck loads would be required to remove the spoil pile.

Problem

A contractor purchases a piece of equipment for $175,000. The equipment's salvage value at the end of 6 years is $17,000. Determine the depreciation for each of these years.

Solution
The formula for calculating depreciation is given below:

$$Depreciation_{n,year} = \left(\frac{Inverse \, year}{Sum \, of \, years}\right) \times Total \, depreciated \, amount$$

The total depreciated amount is $175,000 - $17,000 = $158,000.

The depreciation over each year is:
$$D_1 = \left(\frac{6}{21}\right) \times \$158,000 = \$45,143$$
$$D_2 = \left(\frac{5}{21}\right) \times \$158,000 = \$37,619$$
$$D_3 = \left(\frac{4}{21}\right) \times \$158,000 = \$30,095$$
$$D_4 = \left(\frac{3}{21}\right) \times \$158,000 = \$22,571$$
$$D_5 = \left(\frac{2}{21}\right) \times \$158,000 = \$15,048$$
$$D_6 = \left(\frac{1}{21}\right) \times \$158,000 = \$7,524$$

Check: $45,143 + $37,619 + $30,095 + $22,571 + $15,048 + $7,524 = $158,000

Problem

Calculate the owning and operating costs of a dump truck with an initial cost of $125,000, a salvage value of $20,000, a tire cost of $15,000 (tire life of 3,500 hours). Assume the following: a) insurance, taxes, and storage fees amount to 7% of the average investment, b) hourly depreciation of $15 per hour, c) a fuel consumption rate of $16 per hour, d) an equipment service cost of 33% of the fuel cost, e) an hourly repair cost $12 per hour, f) an investment rate of 8.4%, and g) a plan to operate the dump truck 2,500 hours per year of its life. Disregard operator wages (operating cost) for this problem.

Solution
Owning costs:
Depreciation (given) = $15 per hour
Average investment = (Initial cost + Salvage value)/2 = ($125,000 + $20,000)/2 = $72,500
Hourly investment cost = (0.084 × $72,500)/2500 hours = $2.44 per hour
Hourly insurance, tax and storage cost = (0.07 × $72,500)/2500 hours = $2.03 per hour
Total owning costs = $15/hr + $2.44/hr + $2.03/hr = **$19.47 per hour**

Operating Costs:
Hourly fuel cost (given) = $16 per hour
Hourly service cost = 0.33 × $16/hour = $5.28 per hour
Hourly repair cost (given) = $12 per hour
Hourly tire cost = (1.15 × tire cost)/tire life = (1.15 × $15,000)/3500 hours = $4.93 per hour
Total operating costs = $16 + $5.28 + $12 + $4.93 = **$38.21 per hour**

Methods for construction scheduling

Gantt/bar charts are the simplest of the three when it comes to construction scheduling. A Gantt/bar chart schedule does not show any relationships between the other construction activities, just a start and finish date; each activity is represented as independent of the others. The *critical path method (CPM)* does show relationships between activities, and is typically used for more complicated projects. CPM diagrams graphically show the sequence of events that must occur in order for the shortest overall project completion time. In a CPM diagram, arrows and nodes indicate which activities take precedence over other activities. CPM diagrams may be shown with arrow on activity (AOA) or activity on node (AON) notation. The *program evaluation and review technique (PERT) method* for construction scheduling uses the critical path method along with statistical analysis to determine the likely duration of each activity along the critical path in a project. A standard deviation and variance are calculated for each critical path activity, and from this data, a likely project completion date is determined.

Problem

For the CPM diagram (AOA notation), determine the activities that are on the critical path, the project duration, and the total float. The durations for the activities are shown below:
Activity A: 10 days   Activity B: 7 days
Activity C: 15 days   Activity D: 14 days
Activity E: 11 days   Activity X (dummy): 0 days

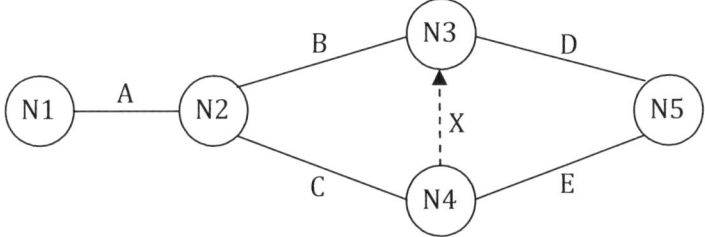

Solution
To calculate the project duration and the total float (TF), the early start (ES), early finish (EF), late start (LS), and late finish (LF) need to be determined for each activity. The arrow X indicates that Activity D cannot begin until Activity C is completed. EF is calculated from EF = ES + Duration. Go backwards to determine LF. LF is equal to the smallest LS of the activity successors. LS is calculated from LS = LF − Duration. TF is calculated from TF = LS − ES. An activity is on the critical path is the total float for that activity is 0.

Activity	Duration (days)	ES	EF	LS	LF	TF	Critical Path Y/N
A	10	0	10	0	10	0	Y
B	7	10	17	18	25	8	N
C	15	10	25	10	25	0	Y
D	14	25	39	25	39	0	Y
E	11	25	36	28	39	3	N

Thus, Activities A, C, and D are the critical path activities, the project duration is 39 days, and the total float for Activities B and E are 8 days and 3 days, respectively.

Problem

Describe what the following CPM diagram (AON notation) illustrates with respect to activity dependence.

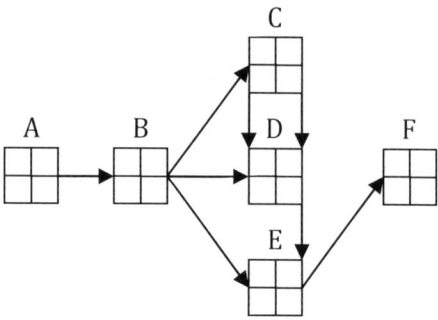

Solution
The CPM diagram shows the following dependencies among the activities:
-The start of Activity B depends on the completion of Activity A.
-The start of Activities C, D, and E depends on the completion of Activity B.
-The start of Activity D depends on the start of Activity C.
-The completion of Activity D depends on the completion of Activity C.
-The completion of Activity E depends on the completion of Activity D.
-The start of Activity F depends on the completion of Activity E.

The critical path for this project would likely include all activities, due to the dependence of activity start and/or completion on the other activities. This type of CPM schedule would be useful in projects where equipment, system components, and software being installed interfaced with each other.

Problem

For the CPM shown, the minimum duration, most likely duration, and maximum duration for Activities A, B, C, and D are given below in days, respectively. Using the PERT, determine the expected duration, the standard deviation, and the variance for each activity. Also determine the critical path.

Activity A: 5, 8, 11 Activity B: 7, 9, 17
Activity C: 5, 6, 13 Activity D: 6, 7, 20

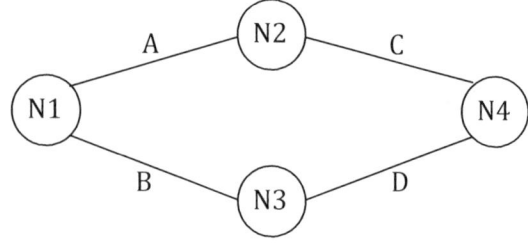

Solution
The expected (mean) activity duration is used to determine the critical path and is calculated using the following equation:

$$t_{expected} = \frac{(t_{min} + 4t_{most\ likely} + t_{max})}{6}$$

For activity A, the equation becomes:
$$t_{A,expected} = \frac{(5 + 4 + 8 + 11)}{6} = 8\ days$$

Similarly solving for Activities B, C, and D yield the following:
$t_{B,expected} = 10\ days, t_{C,expected} = 7\ days$, and $t_{D,expected} = 9\ days$

The critical path is Activities B and D (longest expected duration).
The standard deviation for the activity duration times is calculated from:
$$\sigma = \frac{(t_{max} - t_{min})}{6} \text{ and } \sigma_a = \frac{(11 - 5)}{6} = 1\ day$$

Similarly, $\sigma_B = 1.7\ days, \sigma_C = 1.3\ days$, and $\sigma_D = 2.3\ days$
The variance of the duration times is calculated from the following equation:
$v = \sigma^2$, and $v_A = (1\ day)^2 = 1\ day^2$
Similarly $v_B = 2.9\ days^2, v_C = 1.7\ days^2$, and $v_D = 5.3\ days^2$

Problem

For a CPM, the expected (mean) duration, standard deviation, and variance for the activities on the critical path (Activities A and C) are given, respectively. Using the PERT, determine the critical path's expected duration, variance, standard deviation, and number of standard deviations from the mean (standard normal variable) for completing the project in 14 days.
Activity A: $t_{A,expected} = 7\ days, \sigma_A = 2\ days, v_A = 4\ days^2$
Activity C: $t_{C,expected} = 9\ days, \sigma_C = 5\ days, v_C = 25\ days^2$

Solution
The expected duration of the critical path is the sum of the expected durations of the activities along the critical path:
$$t_{critial\ path,expected} = t_{A,expected} + t_{C,expected} = 7 + 9 = 16\ days$$

The variance of the critical path is the sum of the variances of the activities along the critical path:
$$v_{critical\ path} = v_A + v_C = 4\ days^2 + 25\ days^2 = 29\ days^2$$

The standard deviation of the critical path is calculated from the following equation:
$$\sigma_{critical\ path} = v_{critical\ path}^{1/2} = (29\ days^2)^{1/2} = 5.4\ days$$

The number of the standard deviations from the mean (standard normal variable) for completing the project in 14 days is calculated from the following equation:
$$z = \frac{(t_{interest} - t_{critical\ path,expected})}{\sigma_{critical\ path}} = \frac{(14\ days - 16\ days)}{5.4\ days}$$
$$z = -0.4$$

The number of standard deviations from the mean (standard normal variable), z, is used to determine the probability of completing the project by the duration of interest, a tabulated value.

Construction sequencing

Construction sequencing is the planning of project's construction activities to complete the project in a timely manner to avoid costs or fees incurred due to delays in the project's expected completion date. When determining the construction sequencing, several factors such as availability of land, labor, and resources, project constraints, dependence and precedence of construction activities over other activities, and anticipated weather-related impacts or limitations must be taken into account. Construction sequencing is typically time-driven, with a project completion date in mind. A critical path, the minimum time needed to complete the project, is usually identified and used to track the progress of a project. If the critical path is identified, the contractor is then able to identify the amount that each non-critical activity may be delayed without resulting in a delay of the project (float).

Resource scheduling

Resource scheduling is scheduling a construction project based on the availability of resources, and not necessarily on time. This type of scheduling is necessary when restrictions apply to a project's available resources, such as equipment or labor. Not considering the availability of resources could result in significant delays due to bottlenecking of construction activities dependent on a common resource. If labor is a limiting resource, the contractor may decide to pay for overtime, to open up the resource's availability (time-cost tradeoff example). For limited resources such as equipment, it is crucial that the contractor carefully plan the use of that equipment throughout the entire project. The contractor should have each required use of that equipment identified and planned on a project schedule to ensure the equipment is being used as much as possible with minimum downtime to reduce delays due the equipment's lack of availability.

Costs

Indirect costs are ongoing project costs that are not directly linked to construction activities. Examples of indirect costs include taxes, administrative costs (including administrative staff), office rental fees, and fees associated with office-related services such as telephone, internet, water, and electricity. Indirect costs are typically relatively steady throughout the duration of a project. The shorter the project, the lower the indirect costs will be.

Direct costs are project costs that are directly linked to construction activities. Examples of direct costs include management and staff salaries, travel, construction labor costs, equipment costs, and material costs. Direct costs are not steady throughout the project; they are dependent on the construction activity being performed.

The *project cost* is the sum of indirect costs and direct costs. The actual project cost cannot be determined until after the project is completed.

Time-cost trade-off analysis

Time-cost trade-off analysis is an analysis performed by the contractor to determine which project activities could be compressed or 'crashed' to shorten the duration of the project and reduce overall project costs thereby increasing revenue. Typically, a contractor will focus on critical path construction activities during the analysis since the critical path activities directly relate to project duration and indirect costs. When evaluating potential activities to crash, the contractor should account for available resources (i.e., if enough resources available to complete that activity earlier than originally planned), the costs associated with the crashing compared to the savings, and the level of quality or precision desired and required for an activity. Crashing certain construction activities would risk work quality and potentially become a bigger problem for the contractor in the future. Also, crashed activities can alter the project critical path, so it is imperative the contractor reevaluate the project critical path once activities have been selected for crashing.

Crash cost per day

In evaluating the effects of schedule crashing, the optimum project duration is the point at which the total project costs (indirect and direct) would be at their lowest. When an activity is crashed, the direct costs associated with that activity will increase due to the increase of resources required to complete the activity in a shorter amount of time. The indirect costs for the project, however, will decrease since the overall project duration is shortened. A helpful tool in evaluating activities for crashing potential is determining the crash cost per day of an activity. The daily crash cost is calculated by the following equation:

$$Crash\ cost\ per\ day = \frac{(crash\ cost - normal\ cost)}{(normal\ time - crash\ time)}$$

The activities with the lowest daily crash costs would have the highest potential for successful crashing and should be targeted first, assuming the activities are good candidates for crashing in the first place.

Problem

For the CPM shown, determine which activities should be crashed to achieve a project duration of 16 days, given the following (ND = normal duration, CD = crash duration, NC = Normal Cost, and CC = crash cost):

Activity A:
ND = 8 days, NC = $60,000;
CD = 7 days, CC = $70,000

Activity C:
ND = 10 days, NC = $195,000
CD = 6 days, CC = $240,000

Activity B:
ND = 11 days, NC = $100,000
CD = 10 days, CC = $120,000

Activity D:
ND = 6 days, NC = $170,000
CD = 5 days, CC = $180,000

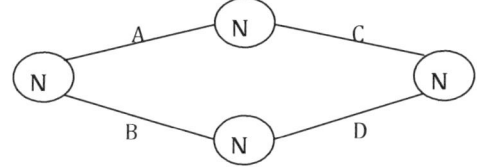

Solution

The duration of Activities A and C is 18 days, and the duration of Activities B and D is 17 days, thus Activities A and C are the critical path activities. Calculate the activity daily crash cost:

$$\text{For Activity A: } \frac{(\$70{,}000 - \$60{,}000)}{(8 \text{ days} - 7 \text{ days})} = \$10{,}000/day$$

Similarly, the daily crash costs for Activities B, C and D are $20,000/day, $11,250/day, and $10,000/day, respectively. Because Activity A's daily crash cost is lower than Activity C's daily crash cost, Activity A should initially be shaved. Activities A and C must be reduced by 2 days, and since Activity A may only be crashed one day, Activity C must also be crashed by one day. The crashed durations for Activity A and C are 7 days and 9 days, respectively (total of 16 days). Now, Activities B and D become critical path with a total duration of 17 days. Since Activity D has a lower daily crash cost compared to Activity B, Activity D should be crashed by one day to achieve the total project duration of 16 days.

Proctor compaction testing

Proctor compaction tests are performed to ultimately determine Optimum Moisture Content (OMC) and the Maximum Dry Density (MDD) of the soil being tested. A soil sample undergoes compaction specific to the test protocol. Upon compaction, the wet density is determined by knowing the weight of the mold, the volume of the mold, and the weight of the soil and the mold. A representative sample from the mold is weighed, then dried and weighed again to obtain the dry weight. From these values, the moisture content can be determined. From the wet density and the moisture content, the dry density of the soil can be determined. Soil compactions are done at five different moisture contents. The dry density is then plotted against the moisture content, and the peak of the resulting curve is the MDD for the soil, and the corresponding moisture content along the x-axis is the soil's OMC.

Problem

Explain the difference between the standard Proctor (AASHTO T99) and the modified Proctor (T180) compaction tests.

Solution

In the Proctor (AASHTO T99) compaction test, a 4-inch diameter mold having a volume of 1/30 cubic foot is used. A 5.5-lb hammer is dropped from 12 inches above the soil sample 25 times. This is performed on three separate but equal lifts of soil in the mold. The compaction produced from this method is approximately 12,400 ft-lb/cf. The modified Proctor (T180) compaction test uses the same mold, but uses a 10-lb hammer dropped 25 times from 18 inches above each of five separate but equal soil lifts instead. The compaction resulting from the modified method is approximately 56,200 ft-lb/cf. Using the modified Proctor compaction test, the resulting MDD is typically higher than the MDD from the standard Proctor test, and the OMC is typically lower than the OMC from the standard Proctor test. The modified Proctor test is typically only used when a higher level of compaction is required in an area of the project.

Concrete strength acceptance testing

Per the American Concrete Institute (ACI)'s Building Code (ACI 318-05), a statistical concrete strength acceptance testing method is available for concrete plants with past testing records for similar materials. The plant must have 30 consecutive test records using a similar material within 1,000 psi of the specified concrete compressive strength (fc') for the proposed material. Two sets of equations are given, one set for compressive strengths less than or equal to 5,000 psi, and another set for compressive strengths greater than 5,000 psi, where s is the standard deviation of the strength tests. From the two equations for a specified compressive strength, the actual required target compressive strength (f_{cr}') is the higher of the two values produced from the set of equations shown below.

For $f_c' \leq 5,000\ psi$: $f_{cr}' = f_c' + 1.34s$; $f_{cr}' = f_c' + 2.33s - 500$
For $f_c' > 5,000\ psi$: $f_{cr}' = f_c' + 1.34s$; $f_{cr}' = 0.9f_c' + 2.33s$

When relevant statistical data from a concrete plant is unavailable, ACI provides a method for concrete strength acceptance testing for concrete specimens cured in a laboratory-type environment meeting ASTM standards. Each test consists of at least two concrete cylinders tested at the same age (usually 28 days). The testing is considered acceptable if the compressive strengths for each test (average cylinder value) are within 500 psi of the design compressive strength for specified strengths less than 5,000 psi (with no limit if the tested compressive strength is higher than the specified strength). For specified strengths greater than 5,000 psi, the compressive strengths for each test must be within 0.10 times the specified strength. The results from any three consecutive averaged tests also must be equal or greater than the specified compressive strength.

Marshall mix hot mix asphalt test method

The main objective of the Marshall mix test method is to determine the optimum asphalt binder content for a mixture given the project design criteria. Before the samples are even tested, an optimum asphalt binder content is estimated. The optimum asphalt binder content determined from the samples must have two asphalt content samples (typically in 0.5% asphalt content increments) below and at least two samples above the optimum; that is why it is necessary for the asphalt binder content to be initially estimated. Four-inch cylindrical samples 2.5 inches in height are prepared in adherence to specific procedures. The samples are compacted by a hammer dropped 18 inches above the sample surface. Upon compaction, the samples then undergo analysis, including a stability and flow test and a density and voids analysis. The optimum asphalt binder content is determined by using the results of the above analyses and looking at their plotted relationships against the percent asphalt content.

Temporary structures

Scaffolding, concrete formwork, falsework and shoring, cofferdams, underpinning, and earth-retaining structures in general are all examples of temporary structures used in construction projects. Temporary structures provide a means of access to work areas during all phases of a construction project to facilitate completion of work in an efficient manner. Temporary structures are vital for employee safety as well. For less complex temporary structures, the general contractor is typically responsible for the design of the structures, whereas, for more complex temporary structures, the design engineer will typically be responsible. Temporary structures can be incorporated into the final project, or removed at the end of their use. Designing a temporary structure is extremely important in that the structure will experience loads the permanent structure may not. Examples of loads that must be taken into account for temporary structures include dynamic or fluctuating loads, impact loads, and shifting loads along with the more permanent load types: live loads, dead loads, and geographic-related loads such as wind, snow and earthquake loads.

Problem

Lateral concrete wall form pressure calculations and assumptions

The lateral wall form pressure calculations are dependent on the temperature, T, of the concrete during its placement in deg F, rate of concrete placement, Rp, in fph, and height of concrete placement, h, in ft. The calculations may be used as long as the following assumptions are met: the temperature of the concrete is between 40 and 90 degrees F, the concrete is Type I cement (containing no admixtures or pozzolans), the density of the concrete is 150 pcf, during placement, the concrete is consolidated via vibration, but not deeper than 4 feet, and the concrete slump is 4 inches or less.

For an Rp 7 feet per hour or less, the lateral wall form pressure, P, in psf, is the least of the following, but at least 600 psf:

$$i)\ P = 150 + \frac{9000 \times R_p}{T}; ii)\ P = 150 \times h; iii)\ P = 2000\ psf$$

For an Rp greater than 7 fph but not greater than 10 fph, the minimum and equations ii) and iii) still apply but i) becomes:

$$i)\ P = 150 + \frac{43,400}{T} + \frac{2800 * R_p}{T}$$

For an Rp greater than 10 fph, only equation ii) applies.

Maximum design load ratings for the various scaffold platform types as per OSHA standards

Per OSHA standards (29 CFR 1926, Subpart L), the minimum safety factor used in designing scaffolding is four; i.e., the scaffolding and all of its components must support four times the maximum design load plus the weight of the scaffold without failing. Light-duty scaffolding has a maximum load of 25 psf. Medium-duty scaffolding has a maximum load of 50 psf. Heavy-duty scaffolding has a maximum load of 75 psf. The maximum loads listed above assume uniform span distribution. One-person scaffolding has a maximum of 250 pounds located at the center of the span. Two-person scaffolding has a maximum of two loads of 250 pounds placed both 18-inches on the right and on the left of the span center. Three-person scaffolding has a maximum of three loads of 250 pounds placed at the center, 18-inches to the right of span center, and 18-inches to the left of span center.

Geotechnical

Problem

AASHTO and Unified (USCS) soil particle size ranges for the four primary components of soil

The four primary components of soil are gravel, sand, silt and clay. The latter two are considered fine-grained soils, and the former two are considered coarse-grained soils. Per the American Association of State Highway and Transportation Officials (AASHTO), the size range of gravel, sand, silt and clay are 2 – 75mm, 0.075 – 2mm, 0.002 – 0.075mm, and 0.001 – 0.002mm, respectively. Per the Unified Soil Classification System (USCS), the size range of gravel, sand, silt, and clay are 2 -100mm, 0.05 – 2mm, 0.002 – 0.05mm, and < 0.002mm, respectively. The particle size distribution for a soil consisting of gravel and sand is determined from a sieve test, a test with performed with sieves with decreasing opening size. The particle size distribution for a soil consisting of silt and clay is determined from a hydrometer test, a test using Stokes Law for the velocity speed of a particle freely falling to calculate particle size (diameter) and density. Tests results from both methods are plotted to obtain a graphical representation of the particle size distribution.

Problem

Determine the AASHTO soil classification for an inorganic soil with the following characteristics:

Sieve Size (mm)	Percent of Sample Retained on Sieve	Misc. Characteristics
<0.025	14	Liquid Limit (LL) = 61
0.025 – 0.038	8	
0.038 – 0.053	15	Plastic Limit (PL) = 30
0.053 – 0.075	45	
>0.075	18	

Solution
The first step is to calculate the percent passing through a 0.075mm (No. 200) sieve size (F_{200}): 14+8+15+45 = 82%. Per the AASHTO soil classification system, a soil is classified as a silt-clay material if the No. 200 percent passing is greater than 35%. The next step is to calculate the plasticity index (PI) from the following equation:

PI = LL – PL = 61 – 30 = 31

For a liquid limit greater than 41 and a plasticity index greater than 11, the soil is classified as A-7-5 or A-7-6. For plasticity indexes 30 or greater, the soil is classified as A-7-5. The group index is then calculated from the following equation:

I_g = (F_{200} – 35)(0.2 +0.005(LL – 40)) + 0.01(F_{200} – 15)(PI – 10)
I_g = (82 – 35)(0.2 +0.005(61 – 40)) + 0.01(82 – 15)(31 – 10)

$I_g = 28.4 \approx 28$

The AASHTO soil classification is A-7-5 (28).

Problem

Determine the AASHTO soil classification for an inorganic soil with the following characteristics:

Sieve Size (mm)	Percent of Sample Retained on Sieve	Misc. Characteristics
<0.025	6	Liquid Limit (LL) = 37
0.025 – 0.075	20	
0.075 – 0.425	15	Plastic Limit (PL) = 20
0.425 – 2.00	45	
>2.00	14	

Solution
First calculate the percent passing through a 0.075mm (No. 200) sieve size (F_{200}): 6 + 20 = 26%. Per the AASHTO soil classification system, a soil is classified as a granular material if the No. 200 percent passing is 35% or less. Because the soil has a No. 200 percent passing greater than 25%, the soil is classified as a type A-2 material. The next step is to calculate the plasticity index (PI) from the following equation:

PI = LL – PL = 37 – 20 = 17

For a liquid limit below 40 and a plasticity index greater than 11, the soil is classified as A-2-6.
For soils classified as A-2-6 or A-2-7, the group index equation is modified as follows:
$I_g = 0.01(F_{200} - 15)(PI - 10)$
$I_g = 0.01(26 - 15)(17 - 10)$
$I_g = 0.77 \approx 1$
The AASHTO soil classification is A-2-6 (1).

Problem

Determine the Unified Soil Classification System (USCS) for an inorganic soil with the following characteristics:

Sieve Size (mm)	Percent of Sample Retained on Sieve	Misc. Characteristics
<0.025	4	Liquid Limit (LL) = 40
0.025 – 0.075	10	
0.075 – 2.00	25	Plastic Limit (PL) = 20
2.00 – 4.75	40	
>4.75	21	

Solution
The USCS classifies soils by using a designated symbol and name for the various soil classes. First, calculate the percent passing through a 0.075mm (No. 200) sieve size: 4 + 10 = 14%. Per the USCS, a soil is classified as a coarse-grained soil if less than 50% of the soil sample passes through a No. 200 sieve. Calculate the percent passing through a 4.75 mm (No. 4) sieve: 4 + 10 + 25 + 40 = 79%. Per the USCS, a soil is classified as a sandy soil if over half of the sample passes through a No. 4 sieve. If over half of the sample did not pass through a No. 4 sieve, the sample would be classified as a gravelly soil. Calculate the plasticity index (PI) from the following equation:

PI = LL − PL = 40 − 20 = 20

Because the percent passing through a No. 200 sieve is 12% or more and the PI is over 7 and above the USCS plasticity chart A-line, the soil is classified as SC: a clayey sand.

Hazen uniformity coefficient and the coefficient of gradation

The Hazen uniformity coefficient, C_u, and the coefficient of gradation, C_z, are two parameters that can be determined from a particle size distribution curve (Percent finer by weight vs. Grain Diameter size, mm, log scale). C_u is calculated from the following equation:

$$C_u = \frac{D_{60}}{D_{10}}$$

Where D_{60} is the diameter at which 60% of the particles are finer by weight, and D_{10} (effective grain size) is the diameter at which 10% of the particles are finer by weight. Soil is considered uniform with regard to particle size if C_u is less than 4 or 5. The coefficient of gradation is calculated from the following equation:

$$C_z = \frac{D_{30}^2}{D_{10} \times D_{60}}$$

Where D_{30} is the diameter at which 30% of the particles are finer by weight.

Problem

Draw a diagram showing the three phases of soil. Give equations for porosity and density.

Solution
Soil contains voids that are either occupied by water or air. Thus soil is a material with three phases: solids, water and air. The mass and volume of the air, voids, water, and solids is broken down below:

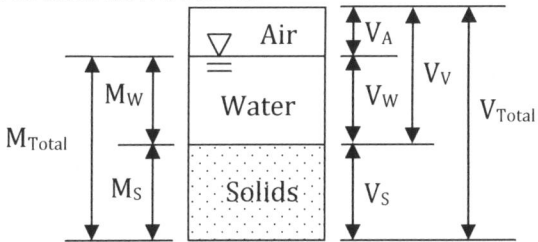

Porosity, n, can be calculated by the following equation:

$$n = \frac{V_V}{V_{Total}} = \frac{V_A + V_W}{V_{Total}}$$

Density, ρ, can be calculated from the following equation:

$$\rho = \frac{M_{Total}}{V_{Total}} = \frac{M_W + M_S}{V_A + V_W + V_S}$$

Problem

A soil sample is sent in for analysis. The volume of the sample is 0.02 cubic meters. The sample mass, M_{Total}, is 40 kg. Determine the degree of saturation if the moisture content, w, is 20%, and the specific gravity of the solids, SG is 2.75.

Solution
To calculate degree of saturation for the given information, the void ratio must first be found. To determine the void ratio, the dry density must be determined. Calculate the dry density, ρ_d, of the sample:

$$\rho_d = \frac{M_{Total}}{(1+w)V_{Total}} = \frac{40\ kg}{(1+0.20)0.02m^3} = 1666.7 kg/m^3$$

Next, calculate the volume of the void ratio, e, from the following equation:

$$\rho_d = \frac{SG * \gamma_w}{1+e}$$

Solving for e, the equation becomes:

$$e = \frac{SG \times \gamma_w}{\rho_d} - 1 = \frac{2.75 \times 1000 kg/m^3}{1666.7 kg/m^3} - 1 = 0.65$$

Finally, calculate the degree of saturation, S, from the following:

$$S = \frac{w \times SG}{e} = \frac{0.20 \times 2.75}{0.65} = 0.85\ or\ 85\%$$

This means that 85% of the void space in the soil is filled with water. Soil is considered fully saturated if the degree of saturation is 100%.

Problem

The volumes of air, water, and solids in a soil being used for a construction project are 100m³, 300m³, and 500m³, respectively. Determine the density of the soil after an extremely heavy rainfall, assuming the mass of solids in the soil is 1.4x10⁶ kg.

Solution

The total mass must be known to calculate the density of the soil. Because air is assumed to have no mass, the only missing component is the mass of the water, M_W. After a heavy rainfall, all of the void spaces will be completely filled with water. The mass of the water will include account for the volume of the water, but also the volume of the voids now filled with water. M_W is then calculated from the following equation:

$$M_W = (V_W + V_A) \times \rho_W = (300 m^3 + 100 m^3) 1000 kg/m^3$$
$$M_W = 4.0 \times 10^5 kg$$

Next, calculate the total mass:

$$M_{Total} = M_W + M_S = 4.0 \times 10^5 kg + 1.4 \times 10^6 kg$$
$$M_{Total} = 1.8 \times 10^6 kg$$

The total volume is calculated from:

$$V_{Total} = V_A + V_W + V_S = 100 m^3 + 300 m^3 + 500 m^3$$
$$V_{Total} = 900 m^3$$

The density is then calculated from:

$$\rho = \frac{M_{Total}}{V_{Total}} = \frac{1.8 \times 10^6 kg}{900 m^3} = 2000 kg/m^3$$

Standard Penetration Test

The Standard Penetration Test (SPT) is an in-situ method used to measure a soil's resistance to penetration (ASTM D-1586). The SPT involves dropping a 140-lb hammer from a distance of 30 inches to drive a split spoon sampler into the ground. The split spoon sampler allows the operator to retrieve a soil bore sample of the material. Resistance is measured by determining the N-value, the number of blows required to penetrate the soil 12 inches. The number of blows required for three separate 6-inch increment values are recorded on a soil bore log with the latter two typically being added together to obtain the N-value. Correction factors pertaining to hammer efficiency, bore hole diameter, and rod length are generally applied to the number of blows to obtain a corrected number of blows. Refusal is attained if 10 consecutive blows do not result in any penetration, if a total of 100 blows is reached, or if 50 blows is required for a 6-inch increment of penetration.

Problem

Fill in the missing information for the SPT bore log.

Blow No.	N-value (blows/ft)	Total Penetration (ft)	Penetration per Blow (ft/blow)
1		0.1	
2		0.3	
3		0.4	
4		0.4	
5		0.5	
6		0.6	
7		0.9	
8		0.9	
9		0.9	
10		1.0	
11		1.3	
12		1.5	

Solution

Penetration per blow is determined by subtracting the current blow total penetration from the previous total penetration. The N-value is determined from the latter two of three 6-inch (0.5 ft) increments of penetration (blows 6-12).

Blow No.	N-value (blows/ft)	Total Penetration (ft)	Penetration per Blow (ft/blow)
1	-	0.1	0.1
2	-	0.3	0.2
3	-	0.4	0.1
4	-	0.4	0
5	-	0.5	0.1
6	-	0.6	0.1
7	-	0.9	0.3
8	-	0.9	0
9	-	0.9	0
10	-	1.0	0.1
11	-	1.3	0.3
12	7	1.5	0.2

Cone Penetrometer Test

The Cone Penetrometer (or Penetration) Test (CPT) is an in-situ method used to obtain various soil properties. A mechanical version of the CPT is available (ASTM D-3441), but most CPTs are now performed using electronic equipment (ASTM D-5778 and D-6067). During a CPT, a penetrometer with a cone-shaped tip is steadily driven into the earth at the rate of approximately 4 ft/minute (20mm /s) using via hydraulic jacking. An electronic CPT, through the use of parameter-specific sensors, can continuously output detailed digitally recorded data such as pore pressure, cone rod inclination, soil resistivity and conductivity, and temperature. Although no sample is obtained using the CPT method, soil types and classifications can be determined by analyzing the data produced from the sensors. A cone bearing (bars) vs. Friction Ratio (%) chart and the Permeability Coefficient (cm/s) as related to soil types are two helpful charts for soil classification.

Problem

During a Proctor test, a soil sample from a construction site is tested six times. The dry densities and water content for each sample is shown in the chart below. Determine the maximum dry density. Also, determine the percent compaction of 0.5 m³ of the same soil having a dry mass of 530 kg.

Test	Dry Density (kg/m³)	Water content (%)
1	1150.7	3
2	1156.4	7.2
3	1160.1	10.1
4	1161.2	11.9
5	1159.3	13.7
6	1157.1	15.1

Solution
The dry densities are plotted against the corresponding water content as shown below:

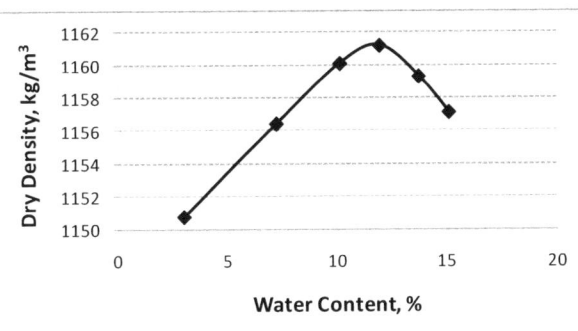

From the graph, the fourth sample has a dry density near the curve peak of the curve, and the maximum dry density is thus ≈1161.2 kg/m³. For the 0.5m³ area of soil, the dry density is:

$$\rho_d = \frac{M_d}{V_{Total}} = \frac{530 kg}{0.5 m^3} = 1060 \ kg/m^3$$

The percent compaction then is: 1060kg/m³ ÷ 1161.2 kg/m³ = 0.91 or 91%

Terms

Atterberg limits are the water contents of a soil during the transitional points between the four potential phases of a soil (solid, semi-solid, plastic solid, and liquid). These limits help distinguish the soil classification.

The *plastic limit* is the water content of the soil associated with the transitional phase between the semi-solid and plastic solid phases. Sands and a majority of silts are considered nonplastic soils and have no plastic limit.

The *liquid limit* is the water content of the soil associated with the transitional phase between the plastic solid and liquid phases.

The *plasticity index* is the difference between the liquid limit and the plasticity limit, and it is an indicator of the moisture content range during the plastic solid phase.

The *liquidity index* is the difference in the water content and plastic limit divided by the difference in the liquid limit and the plastic limit. It is an indicator of the soil consistency.

Problem

During a plastic limit test performed on a soil, the water content results for the three tests were 17.5%, 19% and 17.4%. A liquid limit test was also performed on the soil sample. From the results, the water content during soil rejoining at 25 blows was found to be 31%. Determine or calculate the liquid limit and the plasticity index for this soil.

Solution
During a liquid limit test, the liquid limit is the water content of the soil when it has rejoined for 0.5 inch at precisely 25 blows. Thus the liquid limit is 31%. To calculate the plasticity index, the plastic limit must be determined. The plastic limit (PL) is calculated as an average of the three water content results from the plastic limit test:

$$PL = \frac{17.5\% + 19\% + 17.4\%}{3} = 18\%$$

The plasticity index (PI) is calculated as the difference between the liquid limit (LL) and PL:

$$PI = LL - PL = 31 - 18 = 13$$

Since the plasticity index is not greater than 20, it indicates that a relatively low amount of water introduced to the soil will cause the soil to exhibit liquid qualities.

Problem

Water flowing through a section of soil has a discharge velocity of 4.32 m/hr and a hydraulic gradient of 1.43. If the voids ratio at this corresponding coefficient of permeability is 0.8, provide an estimate of the coefficient of permeability at a voids ratio of 0.6.

Solution
The first step is to convert the units of discharge velocity to match those of typical coefficients of permeability, cm/s.

$$\frac{4.32\ m}{1\ hr} \times \frac{1\ hr}{3600\ s} \times \frac{100\ cm}{1\ m} = 1.2 \times 10^{-1} cm/s$$

The coefficient of permeability, K, can be calculated from the discharge velocity, v, and the hydraulic gradient, i:

$$K = \frac{v}{i} = \frac{1.2 \times 10^{-1} cm/s}{1.43} = 8.4 \times 10^{-2} cm/s$$

The coefficient of permeability indicates that this soil is a fine sand. For granular soils, the coefficient of permeability can be estimated at various void ratios by the following equation:

$$\frac{K_1}{K_2} = \frac{e_1^2}{e_2^2}, \text{ and } K_2 = \frac{K_1 \times e_2^2}{e_1^2} = \frac{8.4 \times 10^{-2} cm/s \times 0.6^2}{0.8^2}$$

$$K_2 = 4.7 \times 10^{-2} cm/s$$

Problem

A constant head permeameter is used to determine the permeability of a soil sample. The length of the soil sample is 6 cm and the diameter of the permeameter is 10 cm. Under a constant head of 60 cm, the discharge volume collected from the permeameter after 60 seconds is 105 cm³. Determine the soil permeability.

Solution

A constant head permeameter is used for sands with larger coefficients of permeability (typically greater than 10^{-4} cm/s), whereas a falling head permeameter is used to test soils with smaller pore openings (such as clays).

The permeability from a constant head permeability test can be calculated from the following equation:

$$K = \frac{VL}{Aht}$$

Where V is the permeameter discharge volume (cubic cm) measured after a certain time t (seconds). A is the cross-sectional area of the permeameter (square cm), and h is the constant head (cm) applied during the test.

The equation becomes:

$$K = \frac{105 cm^3 \times 6 cm}{\pi \left(\frac{10 cm}{2}\right)^2 \times 60 cm * 60 s} = 2.2 \times 10^{-3} cm/s$$

The high permeability value (greater than 10^{-4} cm/s) indicates the soil sample consists of sand.

Asphalt concrete (flexible) pavement properties

Impermeability, stability, vehicular skid and fatigue resistance, flexibility, durability and workability are all desirable traits for asphalt concrete (flexible) pavements. Impermeability of the asphalt concrete to air and water within specified limits is important to maintain the integrity of the pavement. The stability of the asphalt concrete pavement, or the ability of the asphalt concrete to maintain its original condition (smoothness and shape) without any long-term physical damage, is also desirable. Vehicular skid resistance refers to the ability of the flexible pavement to provide enough friction to resist vehicular skidding or sliding, and fatigue resistance refers to its ability to endure continual vehicular-related loads without compromise. A flexible pavement's durability refers to weather-related pavement disintegration. The flexibility of asphalt concrete pavement refers to the ability of the pavement to withstand small shifts or movements in the subgrade without cracking. Workability refers to the effort required in placing and compacting the flexible pavement.

Components of asphalt concrete pavement

Mineral aggregate and asphalt cement (asphalt binder) are the two main components of asphalt concrete pavement. Mineral aggregate is made up of crushed rock, gravel, or sand and accounts for approximately 90-95% of the asphalt mixture by weight. Precision of the aggregate grading is important. Aggregate size range restrictions exist for pertinent sieve opening sizes. The allowable lift thickness also depends on the size of the largest aggregate. The asphalt binder is a sticky, tar-like substance, produced typically in petroleum refineries, that acts as a glue to bind the mineral aggregates together. Since the durability and reliability of the binder to bind the aggregates is vital to the flexible pavement's life, an asphalt modifier may be added to the mix to help ensure a longer life for the flexible pavement. Examples of asphalt modifier types include polymers, rubbers, plastics, and mineral fillers.

Problem

Determine the percent of a) voids in the total mix, b) voids in the mineral aggregate, and c) voids filled with asphalt for a mixture having the following characteristics. Assume a compacted mixture bulk specific gravity, G_{sb} of 2.71.
Test Sample Maximum Asphalt Content, $(100\% - P_s) = 6\%$
Test Sample Asphalt specific gravity, $G_{mm} = 2.40$
Desired specific gravity, $G_{mb} = 2.35$

Solution
a) To calculate the percent of voids in the total mix, VTM, use the following equation:

$$VTM = \left(1 - \frac{G_{mb}}{G_{mm}}\right) \times 100\%$$
$$VTM = \left(1 - \frac{2.35}{2.40}\right) \times 100\% = 2.08\%$$

b) The percent of voids in the mineral aggregate, VMA, may be calculated from:

$$VMA = 100\% - \frac{G_{mb} P_s}{G_{sb}}$$
$$VMA = 100\% - \frac{2.35 \times 94\%}{2.71} = 18.5\%$$

c) The percent of voids filled with asphalt, VFA, is calculated from:

$$VFA = \left(1 - \frac{VTM}{VMA}\right) \times 100\%$$
$$VFA = \left(1 - \frac{2.08\%}{18.5\%}\right) \times 100\% = 88.8\%$$

*A VTM below 3% indicates that this soil will not have adequate room for binder expansion during hot weather.

Flexible pavement design

AASHTO (1993) method
The AASHTO (1993) method for flexible pavement design ultimately yields the minimum required pavement layer thicknesses to withstand the predicted traffic loads. The AASHTO method requires that the equivalent single-axle load (ESAL) factor be determined for future conditions (based on 18,000 pound single-axle single pass loads). This factor is used to estimate the amount of damage caused by the loads over the pavement. The basic AASHTO method provides an empirical formula to determine the structural number (SN) needed to withstand design ESALs greater than 50,000. An SN of 5 can be initially assumed, but would need to be verified prior to design finalization. Two optional steps of the basic AASHTO method include consideration for staged construction and for roadbed swelling and frost heave. In the final step, the pavement layer thicknesses (subbase, base, and surfacing) are calculated from the design SN. Since the layer thicknesses are calculated simultaneously, many combinations of layer thicknesses would suffice, but AASHTO suggests criteria to follow when determining the layer thicknesses.

Asphalt Institute (1991) method
The Asphalt Institute (AI) (1991) method for flexible pavement design determines the minimum layer thickness that will resist critical horizontal tensile strains at the asphalt layer bottom and critical vertical compressive strains at the subgrade surface. With this method, the ESAL factor for 18,000 pound axle single pass loads is estimated for future conditions. The subgrade resilient modulus is then determined by testing samples of the subgrade material to develop a smooth S-curve of the resulting resilient moduli (by percent of values equal to or greater than test value). A resilient moduli percentile vs. ESAL range chart is used to determine the percentile at which the design resilient modulus value can be found on the curve. The base types to be analyzed are identified and a design thickness for each type is determined from provided design charts for hot, cold, and mild temperatures. This method includes optional consideration for staged construction. An economic analysis comparing the design alternatives is then performed, and a final design is selected.

Rigid pavement design

AASHTO (1993) method
The AASHTO (1993) method for rigid pavement design ultimately yields the minimum required slab thickness to withstand the predicted traffic loads. The AASHTO method requires that the equivalent single-axle load (ESAL) factor for 18,000 pound axle single pass loads be determined for future conditions. For design ESALs greater than 50,000, the following applies. The effective modulus of subgrade reaction, k, is determined through a series of steps involving the use of nomographs. In the first step of this method, various levels of slab support are considered, producing a k-value for each combination considered. As with flexible design, two optional steps of the rigid pavement design method include

consideration for staged construction and for roadbed swelling and frost heave. The required slab thicknesses can be obtained for each k-value using an empirical formula or nomograph. Upon evaluation of the results, method guidelines, and design conditions, a design k-value and corresponding slab thickness is selected.

Portland Cement Association (1984) method

The Portland Cement Association (PCA) (1984) method for designing rigid pavement considers fatigue and erosion analysis. This method can be used with or without axle load data. For analyses with available axle load data, the PCA method only takes into account vehicles with six or more axles. In addition to axle load data, other factors taken into account with this method include a trial pavement thickness, joint type, shoulder type, concrete modulus of rupture, k-value of subgrade (including subbase), load safety factor, and axle load distribution. From these date inputs and with the aid of provided charts and nomographs, the fatigue and erosion factors can be determined. The allowable repetitions are determined separately for the fatigue and erosion factors, and a total percent of fatigue damage and erosion damage is then calculated. The process is iterative, in that if the damage due to fatigue or erosion is greater than 100%, a new trial thickness must be selected and the method repeated, until that condition is met.

Problem

Using the AASHTO flexible pavement method, calculate the truck factor for 103 total single-axle trucks using the axle load information below. Assume a terminal serviceability index, p_t, of 2.5 and a SN of 4.

Axle Load (kips)	Axle Type	No. of Passes
10	Single	45
18	Single	58
20	Tandem	110
32	Tandem	71
52	Triple	32

Solution

AASHTO provides tabulated load equivalency factors (LEFs) for single, tandem, and triple axle loads with p_t values of 2.5 and *SN* values 1 through 6. The ESAL is calculated by multiplying the number of passes by the LEF for each axle load.

Axle Load (kips)	Axle Type	No. of Passes	LEF	ESAL
10	Single	45	0.102	4.590
18	Single	58	1.00	58.000
20	Tandem	110	0.141	15.510
32	Tandem	71	0.887	62.977
52	Triple	32	1.44	46.080
			Total ESALs	187.157

The sum of ESALs is then divided by the number of trucks to obtain the truck factor, TF, for a given class: TF = 187.157/103 = 1.82 ESAL/truck.

Contraction joints and construction joints

Contraction joints and construction joints are common types of rigid pavement joints. Contraction joints, or control joints, are purposed joints either sawn after concrete has hardened or hand-formed during concrete placing. The joint causes a reduction in pavement thickness typically between one- fourth to one-third of the total slab depth. These joints create a weak point in the pavement and help reduce slab cracking. Contraction joints may be perpendicular to traffic flow or made at skewed angles. The joint may be skewed so that one front tire of a vehicle passes over the joint before the other front tire (load transfer). Dowels may or may not be used with contraction joints. Construction joints are joints that formed between concrete pours and are either longitudinal or transverse for roadways. Longitudinal joints are typically located between lanes, and transverse joints are based on the concrete available during the pour. Dowels may or may not be used with construction joints.

Expansion joints

Expansion joints, also called isolation joints, are a common type of rigid pavement joints. Expansion joints are located at the point where concrete pavement and a structure meet. Examples of expansion joint applications would be where a slab meets a building foundation, manhole, bridge, or structure footing. Expansion joints are also found at pavement intersections. This type of joint alleviates compressive stresses at these locations and allows for thermal contraction and expansion. The joint material is typically made of impregnated foam or polyurethane compounds, and is fixed to the concrete formwork prior to the pour. The material is placed between the slab and structure and expands and contracts as needed. A joint filler material or sealer compound may be used between the joint material and slab surface to ensure that the joint is protected.

Problem

A rectangular footing is placed over homogeneous soil. The point load, Q, of the 4-ft × 6-ft footing is 6,000 pounds. Using the Boussinesq equation, calculate the increase in vertical pressure at Point A.

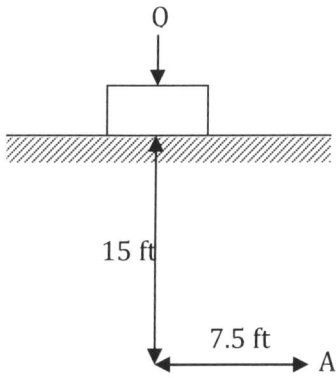

Solution

The Boussinesq equation assumes that the soil is elastic, semi-infinite, isotropic, and homogeneous. The width of the footing is also assumed to be much smaller than the depth to the point of interest (i.e., depth to point of interest must be greater than two times the footing width). In this case, the depth is 15 ft, which is greater than two times the footing width, so this equation could be used. The Boussinesq equation is shown below:

$$\Delta p_v = \frac{3Qh^3}{2\pi(h^2 + r^2)^{\frac{5}{2}}}$$

Where Δp_v is the increase in vertical pressure, Q is the surface point load, h is the vertical distance, or depth, to the point of interest (Point A), and r is the horizontal distance to the point of interest (Point A).

The equation becomes:

$$\Delta p_v = \frac{3 \times 6000lb \times (15ft)^3}{2\pi((15ft)^2 + (7.5ft)^2)^{\frac{5}{2}}} = 7.3 \ psf$$

Problem

Use the 2:1 zone of influence method to determine the increase in vertical pressure of homogeneous soil 15 feet below (Point A) a 4-ft × 6-ft footing with a point load, Q, of 6,000 pounds.

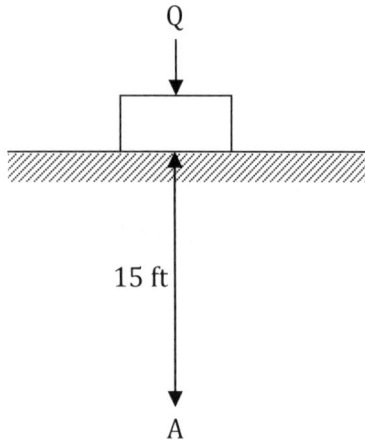

The 2:1 zone of influent method approximates the increase in vertical pressure (stress) at a depth by looking at the geometry (zone of influence). The method also assumes the soil consists of non-layered homogeneous material. For this method to be used, it is necessary that the ratio of the depth to the point of interest to the width of the footing be greater than 1.5 but less than 5. For this example, the ratio is 15 ft/4 ft = 3.75 which falls within the requirement. The 2:1 zone of influent equation is shown below:

$$\Delta p_v = \frac{Q}{(B+h)(L+h)}$$

Where Δp_v is the increase in vertical pressure, Q is the surface point load, h is the vertical distance, or depth, to the point of interest (Point A), B is the footing width, and L is the footing length.

The equation becomes:

$$\Delta p_v = \frac{6000\ lb}{(4ft + 15ft)(6ft + 15ft)} = 15.0 psf$$

Types of lateral earth pressure on a retaining wall

The three types of lateral earth pressure on a retaining wall are at rest, active, and passive. For the at *rest* lateral earth pressure, there is no movement of the retaining wall away from or towards the backfill the wall is in contact with. For conditions with *active* lateral earth pressure (also referred to as tensioned or forward earth pressure), the retaining wall moves away from the backfill. As the wall moves away from the soil, the lateral earth pressure will decrease until the minimum active lateral pressure is reached. For conditions with *passive* lateral earth pressure (also referred to as backward or compressed earth pressure), the retaining wall moves toward the backfill. As the wall moves toward the backfill, the lateral earth pressure increases, until a maximum passive lateral pressure is reached.

Problem

Calculate the at rest total lateral force per the unit length of an 8-ft wall retaining dry granular backfill with the following conditions:
Dry unit weight of soil = 95 lb/ft³
Soil angle of internal friction = 35°

Solution
When the lateral earth pressure of backfill against a retaining wall is at rest, the total lateral force per unit of wall length, P_o, can be calculated from the following equation:

$$P_o = \frac{1}{2} K_o \gamma_d H^2$$

Where K_o is the coefficient of the earth pressure at rest, γ_d is the soil's dry unit weight, and H is the height of the retaining wall. The coefficient of the earth pressure is dependent on the soil angle of internal friction, ϕ, by the following equation:

$$K_o = 1 - sin\phi$$

For this example:

$$K_o = 1 - \sin 35 = 0.43$$
$$P_o = \frac{1}{2}(0.43)\left(95 \, \frac{lb}{ft^3}\right)(8\, ft)^2 = 1307 \, lb/ft$$

Problem

Using the Rankine theory, calculate the active total lateral force per the unit length of an 8-ft wall retaining granular backfill with the following conditions:
The backfill is saturated 4 feet below the top of the retaining wall.
Unit weight of soil, γ = 97 lb/ft³
Saturated unit weight of the soil, γ_{sat} = 110 lb/ft³
Assume the coefficient of the active earth pressure, K_a is 0.31.

Solution
For a granular backfill, the Rankine active pressure, σ_a', and the total Rankine active pressure, P_a, equations are:

$$\sigma_a' = K_a \sigma_v' \quad \text{and} \quad P_a = \frac{1}{2}(\sigma_a')H$$

Where and σ_v' is the effective vertical pressure at a given depth ($\gamma \times H$). The active pressure of the soil near the top of the wall is equal to zero. Four feet below the top, there is no hydrostatic pressure (top of water level) and the active pressure is:

$$\sigma_a' = 0.31 \left(\frac{97 \, lb}{ft^3}\right)(4ft) = 120.3 \, lb/ft^2$$

Eight feet below the top (wall bottom), the hydrostatic pressure is equal to (62.4 lb/ft³)(4ft) = 249.6 lb/ft². The active pressure is:

$$\sigma_a' = 0.31\left(\frac{97\,lb}{ft^3} + \left(\frac{110\,lb}{ft^3} - \frac{62.4\,lb}{ft^3}\right)\right)(4ft) = 179.3 \, lb/ft^2$$

$$\text{Thus, } \boldsymbol{P_a} = \frac{1}{2}\left(\frac{120.3\,lb}{ft^2}\right)4ft + \frac{1}{2}\left(\frac{120.3\,lb}{ft^2} + \frac{179.3\,lb}{ft^2}\right)4ft + \frac{1}{2}\left(\frac{249.6\,lb}{ft^2}\right)4ft = \boldsymbol{1339\, lb/ft}$$

Soil consolidation

Soil consolidation is the process by which the soil void fraction decreases, thereby decreasing the soil volume. There are three different phases of consolidation that may occur. Immediate consolidation or settling occurs immediately after the soil experiences stress from loads on the soil. Primary consolidation refers to the release of water from clay soils while under stress. This gradual release occurs over a long period of time, but decreases with time. Secondary consolidation refers to any soil consolidation that occurs after the completion of primary consolidation. Secondary consolidation may or may not occur, since its occurrence depends on the soil characteristics. A soil will rebound and gain some of its volume back if the stress causing the consolidation is relieved. Normally consolidated clay refers to soils that have experienced the maximum stress known to be applied to that soil type. Overconsolidated or preconsolidated clay refers to soil that has experienced a greater stress in the past compared to current conditions.

Problem

A normally consolidated saturated 6-ft deep clay soil layer has the following characteristics:
Water content, w = 30%
Specific gravity, SG = 2.71
Compression index, C_c = 0.39
Effective overburden pressure at layer center, p_0 = 998 lb/ft²
Increase in vertical pressure, Δp_v = 164.3 lb/ft²

Determine the primary consolidation settlement for the soil.

Solution
The equation to determine primary consolidation settlement for a normally consolidated clay soil, S_c, is:
$$S_c = \frac{C_c H}{1 + e_i} \log \frac{p_0 + \Delta p_v}{p_0}$$
Where H is the saturated clay layer thickness and e_i is the initial void ratio of the clay. The initial void ratio can be calculated from:
$$e_i = \frac{w \times SG}{S}$$
Where S is the degree of saturation. For saturated soils, the degree of saturation is equal to 1.
$$e_i = \frac{0.30 \times 2.71}{1} = 0.81$$
$$S_c = \frac{0.39 \times 6ft}{1 + 0.81} \log \frac{\frac{998 lb}{ft^2} + \frac{164.3 lb}{ft^2}}{\frac{998 lb}{ft^2}}$$

$$S_c = 0.086 ft = 1.03 in$$

Problem

A 9-ft clay layer has a coefficient of consolidation, C_v, of 0.15 ft/day. Determine the time it would take to achieve 70% primary consolidation if the layer has one-way drainage compared to two-way drainage.

Solution
The equation for calculating the time of primary consolidation in a clay layer, t, is:

$$t = \frac{T_v H_d^2}{C_v}$$

Where T_v is a non-dimensional time factor dependent on the average degree of consolidation, U. H_d is the length of the drainage path, and C_v is the coefficient of consolidation. For a degree of consolidation of 70%, the corresponding tabulated value of T_v is 0.403.

For one-way drainage, $H_d = H$, so that:

$$t = \frac{T_v H^2}{C_v} = \frac{0.403 \times (9ft)^2}{0.15\ ft/day} = 217.6\ days$$

For two-way drainage, $H_d = H/2$, so that:

$$t = \frac{T_v H^2}{C_v} = \frac{0.403 \times \left(\frac{9ft}{2}\right)^2}{0.15\ ft/day} = 54.4\ days$$

Problem

A 5-foot surface clay layer exists above an 11-ft sand layer. The water table exists 2 feet below the surface. Determine the effective stress at the surface, at the water table, at the bottom of the clay layer, and at the bottom of the sand layer. The layers have the following characteristics:

Unit weight of clay (below and above water table), γ_{clay} = 115 lb/ft³
Unit weight of saturated sand, γ_{sand} = 130 lb/ft³

Solution
The effective stress is zero at the soil surface, since there are no vertical forces applied at the surface. At the water table, 2 feet below the surface, the effective stress, σ', is:

$$\sigma'_{2ft} = \gamma_{clay} \times H$$

Where H is the depth to the point of interest. At the water table, $H = 2$, so:

$$\sigma'_{2ft} = 115\frac{lb}{ft^3} \times 2ft = 230\ lb/ft^2$$

At the bottom of the clay layer:

$$\sigma'_{5ft} = 320\frac{lb}{ft^2} + \left(115\frac{lb}{ft^3} - 62.4\frac{lb}{ft^3}\right)3ft = 477.8\ lb/ft^2$$

At the bottom of the sand layer:

$$\sigma'_{16ft} = 477.8\frac{lb}{ft^2} + \left(130\frac{lb}{ft^3} - 62.4\frac{lb}{ft^3}\right)11ft$$
$$\sigma'_{16ft} = 1221.4\ lb/ft^2$$

Problem

A 5-foot surface clay layer exists above an 11-ft sand layer. The water table exists 2 feet below the surface. Given the effective stresses below, determine the pore pressure and total stress at the water table, at the bottom of the sand layer, and at the bottom of the clay layer.

σ'_{2ft} = 230 lb/ft²
σ'_{5ft} = 477.8 lb/ft²
σ'_{16ft} = 1221.4 lb/ft²

Solution

The pore water pressure at and above a water table is zero. Below the water table, the pore pressure, u, can be calculated from:

$$u = \gamma_w * H$$

Where γ_w is the unit weight of water, and H is the depth to the point of interest. At the bottom of the clay layer:

$$u_{5ft} = 62.4 \frac{lb}{ft^3} \times (5\ ft - 2\ ft) = 187.2\ lb/ft^2$$

At the bottom of the sand layer:

$$u_{16ft} = 62.4 \frac{lb}{ft^3} \times (16\ ft - 2\ ft) = 873.6\ lb/ft^2$$

The total pressure, σ, is simply:

$$\sigma = u + \sigma'$$

At the water table, since the pore water pressure is zero, the total stress is equal to the effective stress (230 lb/ft²). At the bottom of the clay layer, the total stress is 187.2 lb/ft² + 477.8 lb/ft² = 665 lb/ft². At the bottom of the sand layer, the total stress is 873.6 lb/ft² + 1221.4 lb/ft² = 2095 lb/ft².

Problem

Sketch slope circle diagrams for the three major types of soil slope failure.

Solution

Slope failure is failure in which the sliding surface of the soil intersects the slope at a distance of z above the toe of the slope:

Toe failure is failure in which the sliding surface of the soil intersects the slope at the toe of the slope (z=0):

Base failure is failure at which the sliding surface of the soil occurs a distance of z below the toe of the slope:

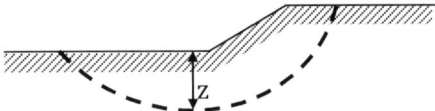

Problem

Determine the cohesive factor of safety to protect against slope failure for saturated clay with the following characteristics:
Unit weight of saturated clay = 120 lb/ft³
Cohesion of saturated clay = 415 lb/ft²
Stability number (from Taylor Slope Stability Chart) = 8.65
Slope height = 45 ft

Solution
The equation for calculating the cohesive factor of safety for saturated clay is:
$$F_{cohesive} = \frac{N_o \times c_{clay}}{H \times \gamma_{eff}}$$
Where N_o is the stability number (from Taylor Slope Stability Chart), c_{clay} is the saturated clay cohesion value, H is the slope height (or depth of cut), and γ_{eff} is the effective specific weight of the clay. The effective specific weight of clay can be calculated from:

$$\gamma_{eff} = \gamma_{sat.\ clay} - \gamma_{water}$$

$$\gamma_{eff} = 120\frac{lb}{ft^3} - 62.4\frac{lb}{ft^3} = 57.6\frac{lb}{ft^3}$$

The cohesive factor of safety is then:
$$F_{cohesive} = \frac{8.65 \times 415\ lb/ft^2}{45ft \times 57\ lb/ft^3} = 1.4$$

A cohesive factor of safety of 1.4 falls within the range of an acceptable minimum factor of safety (typically 1.3 to 1.5).

Problem

A 24-inch concrete pipe is placed in an excavated trench that is 17 feet deep (from ground surface to bottom of pipe) and then backfilled with sand and gravel (unit weight of 100 lb/ft³). The trench width at the top of the pipe is 5 feet. Calculate the dead load (per unit length) on the pipe imposed by the backfill using Marston's formula. Also, calculate what the dead load would be if the pipe is made of plastic.

Solution
Marston's formula for calculating the dead load per unit of length, W, for a rigid pipe (i.e., concrete) is:

$$W = C_{load}\, \gamma B^2$$

Where C_{load} is the buried pipe loading coefficient, γ is the unit weight of the backfill, and B is the trench width at the top of pipe. C_{load} is a tabulated value dependent on the backfill material and the ratio of trench height (surface to top of pipe) to width (at top of pipe). The trench depth to top of pipe is 17 ft – 2 ft = 15 ft. The ratio of trench depth to width is 15 ft/5 ft = 3. The corresponding C_{load} factor for sand and gravel with a unit weight of 100 lb/ft³ and a trench depth to width ratio of 3 is 1.90.

$$W = 1.90 \times 100\, \frac{lb}{ft^3} \times (5 ft)^2 = 4750\, \frac{lb}{ft}\ (concrete\ pipe)$$

If the pipe is flexible (i.e., plastic), Marston's formula becomes:

$$W = C_{load}\, \gamma B D$$

Where D is the pipe diameter. Thus,

$$W = 1.90 \times 100\, \frac{lb}{ft^3} \times 5\ ft \times 2\ ft = 1900\, \frac{lb}{ft}\ (plastic\ pipe)$$

Impacts a soil can have on a concrete slab placed on grade

A slab-on-grade is a shallow foundation that transfers a load to the soil beneath it. The load is transferred to soil near the ground surface, thus the soil directly beneath the slab-on-grade can play a role in the integrity of the slab and structure over time. Clay soils will expand with water (swell) or consolidate (shrink by release of water) over time. Clays are not the preferred type of soil beneath slabs-on-grade for this reason. If a slab is placed over clay with high water content, during dry periods and over time, the soil beneath the slab will consolidate and settle which may cause substantial damage to the concrete slab and structure above. Sands are preferred over clays for bearing shallow foundations. Sand drains relatively quickly compared to clay and if well-compacted, sand will not settle much after its initial loading settlement. Since most soils are not purely clay or sand, analysis must be performed to determine the soil's bearing capacity.

Problem

Determine the skin friction capacity of a driven 14-inch square, 60-ft long concrete foundation pile, given the tabulated information below. Use the α-method (empirical adhesion factor method).

Clay Layer Thickness, ft	Undrained Shear Strength, lb/ft²	Empirical Adhesion Factor
15	1000	0.85
20	2000	0.49
25	3000	0.52

Solution
The skin friction capacity, Q_f, can be calculated from the equation below:
$$Q_f = p \sum \alpha C_u \Delta L$$
Where p is the perimeter of the pile cross-section, α is the empirical adhesion factor, C_u is the undrained shear strength, and ΔL is the soil layer thickness.
The perimeter of the pile cross-section is:
$$p = \frac{14 \text{ in} \times 4}{12 \text{ in/ft}} = 4.67 \text{ ft}$$

Solving for the other summation of $\alpha C_u \Delta L$:

ΔL, ft	Cu, lb/ft²	α	αCuΔL, lb/ft
15	1000	0.85	12,750
20	2000	0.49	19,600
25	3000	0.52	39,000
		Total Sum	71,350

Thus, the skin friction capacity is:
$$Q_f = 4.67 \text{ ft} \times 71,350 \frac{lb}{ft} = 333,205 \text{ lbs} = 333 \text{ kips}$$

Problem

Determine the ultimate static bearing capacity and the allowable capacity of a driven 12-inch square, 60-ft long concrete foundation pile given the information below:
Cohesive clay soil with soil angle of internal friction, $\phi = 0°$.
Skin friction capacity of driven pile = 333 kips.
Undrained shear strength of deepest layer = 3000 lb/ft².
Factor of Safety = 3

Solution
The ultimate static bearing capacity, Q_{ult}, for a driven pile is:
$$Q_{ult} = Q_p + Q_f$$
Where Q_p is the point capacity, and Q_f is the skin friction capacity.
The point capacity is calculated from the following equation:

$$Q_p = A_p \left(\frac{\gamma B N_\gamma}{2} + C_u N_c + \gamma D_f N_q \right)$$

For small pile sizes, the ½(γBN_γ) term is typically omitted. For cohesive soils with an angle of internal friction of 0°, N_q = 1, and the γDf term is considered cancelled by the weight of the pile. For driven piles of conventional dimensions, N_c = 9. Thus, the equation becomes:

$$Q_p = A_p(C_u \times 9) = (1ft \times 1ft)\left(\frac{3000\ lb}{ft^2}\right)9 = 27,000\ lb/ft$$

$Q_p = 27\ kips$

$Q_{ult} = Q_p + Q_f = 27\ kips + 333\ kips = 360\ kips$

The allowable capacity, Q_a, is calculated from:

$$Q_a = \frac{Q_{ult}}{Factor\ of\ Safety} = \frac{360\ kips}{3} = 120\ kips$$

Flow net

A flow net is used to calculate the flow rate of water through soil. A flow net consists of flow lines and equipotential lines. Flow lines (or velocity lines) lines are lines that indicate the path of a water particle through the soil, upstream to downstream. Equipotential lines (or equal head lines) are lines that mark the points in the flow net that have the same potential head. Flow lines and equipotential lines cross each other at right angles. The equation for calculating the flow per unit length, q, is:

$$q = K \times H \frac{N_f}{N_d}$$

Where K is the coefficient of permeability, H is the difference in water levels, N_f is the number of flow channels in the flow net, and N_d is the number of drops in the flow net.

Problem

A water barrier separates two areas (Sides A and B) of a flooded region. The depth of water above grade on Side A is 20 feet. The depth of water above grade on Side B is 2 feet. Calculate the flow per unit length between two sides of this system, given the flow net information below:

Coefficient of Permeability = 0.092 ft/min
Number of Flow Channels = 5
Number of Drops = 6

Solution

$$q = 0.092 \frac{ft}{min} \times (20ft - 2ft)\frac{5}{6} = 1.38\ ft^3/day/ft$$

Liquefaction

Liquefaction is the occurrence in which the shear strength of a saturated cohesion-less soil is suddenly reduced. Liquefaction may occur as a result of rapid loading or earthquake activity. The water between the soil particles increases in pressure to such an extent as to loosen the soil particles, thereby weakening the strength of the soil. When the shear strength of the soil becomes zero, the soil behaves more like a liquid and can cause substantial damage to nearby buildings and structures. A sandy soil may be rated according to its potential for liquefaction to occur. This rating is known as the cyclic stress ratio. It is a ratio of the average cyclic shear stress to the vertical effective stress upon the sand (prior to the sudden loading or earthquake). The ratio and subsequent rating is obtained from laboratory or field testing data.

Problem

Using the Mohr-Coulomb failure criteria equation, determine the shear strength of a soil for the two following conditions. Assume water content remains unchanged. Also determine the unconfined compressive strength of the Condition B soil.
Condition A: Sand
Effective normal stress = 350 kPa
Drained angle of internal friction = 30°
Condition B: Overconsolidated clay
Undrained shear strength (cohesion) = 750 lb/ft²

Solution
The Mohr-Coulomb failure criteria equation to calculate the shear strength of a soil, s, is:

$$s = C + \sigma' \tan\phi$$

Where C is the cohesion of the soil, σ' is the effective normal stress, and ϕ is the drained angle of internal friction. Condition A consists of a sandy soil. For sands and normally consolidated clays, $C = 0$, so:

$$s = \sigma' \tan\phi = 350\ kPa\ (\tan 30) = 202\ kPa$$

For over consolidated clays (cohesion is present), the drained angle of internal friction is 0°, so:

$$s = C_u = 750\ lb/ft^2$$

The unconfined compressive strength, q_u, of a cohesive soil is given by:

$$q_u = 2 * C_u = 2 \times \frac{750\ lb}{ft^2} = 1500\ lb/ft^2$$

Problem

Give the ultimate bearing capacity equation for shallow foundations in clay.

Solution
The general ultimate bearing capacity equation for shallow foundations is:

$$q_{ult} = cF_{cs}F_{cd}N_c + (p_q + \gamma D_f)F_{qs}F_{qd}N_q + \frac{1}{2}\gamma B F_{\gamma s}F_{\gamma d}N_\gamma$$

Where c is the cohesion, p_q is any additional surface surcharge, γ is the unit weight of the soil, D_f is the foundation depth, B is the foundation width, F_{cs}, F_{qs}, and $F_{\gamma s}$ are shape factors, F_{cd}, F_{qd}, and $F_{\gamma d}$ are depth factors, and N_c, N_q, and N_γ are bearing capacity factors.
For clay soils, the angle of internal friction is considered to be 0°. From the Terzaghi Bearing Capacity Factor Chart for General Shear, $N_c = 5.7$, $N_q = 1.0$, and $N_\gamma = 0$ at an angle of internal friction of 0°.

Thus the ultimate bearing capacity equation for shallow foundations in clay becomes:
$$q_{ult} = cF_{cs}F_{cd} \times 5.7 + (p_q + \gamma D_f)F_{qs}F_{qd}$$

Problem

Give the ultimate bearing capacity equation for shallow foundations in ideal sand.

Solution
The general ultimate bearing capacity equation for shallow foundations is:
$$q_{ult} = cF_{cs}F_{cd}N_c + (p_q + \gamma D_f)F_{qs}F_{qd}N_q + \frac{1}{2}\gamma B F_{\gamma s}F_{\gamma d}N_\gamma$$

Where c is the cohesion, p_q is any additional surface surcharge, γ is the unit weight of the soil, D_f is the foundation depth, B is the foundation width, F_{cs}, F_{qs}, and $F_{\gamma s}$ are shape factors, F_{cd}, F_{qd}, and $F_{\gamma d}$ are depth factors, and N_c, N_q, and N_γ are bearing capacity factors.
For ideal sand, cohesion is equal to zero. Thus the ultimate bearing capacity equation for shallow foundations in ideal sand becomes:
$$q_{ult} = (p_q + \gamma D_f)F_{qs}F_{qd}N_q + \frac{1}{2}\gamma B F_{\gamma s}F_{\gamma d}N_\gamma$$

The first term (the depth term) has a substantial effect on the bearing capacity for cohesion-less soils such as sand.

Problem

Use the Schmertmann Method to calculate the total settlement of sand under a 4-ft wide footing, given the following assumptions and the tabulated data for the layers beneath the footing.

Embedment correction factor, $C_1 = 0.977$
Creep correction factor, $C_2 = 1.31$; Net foundation pressure increase, $\Delta p = 2.85$ tsf

Layer Depth (in)	Penetration Resistance, N	Es/N	Ave. Influence Strain Factor, I_z
36	9	4	0.35
24	15	7	0.58
36	24	10	0.22

Solution
The Schmertmann Method equation for calculating total settlement, S, is:
$$S = C_1 C_2 \Delta p \sum_0^{2B} \frac{I_z}{E_S} \Delta Z$$
The E_S term is calculated from multiplying the penetration resistance by the E_S/N term. The summation term can then be calculated from the data provided, where ΔZ is the layer thickness in inches:

Layer Depth (in)	Penetration Resistance, N	E_S/N	Ave. Influence Strain Factor, I_Z	E_S (tsf)	$\dfrac{I_Z}{E_S}\Delta Z$ (in/tsf)
36	9	4	0.35	36	0.350
24	15	7	0.58	105	0.133
36	24	10	0.22	240	0.033
				Total Summation	0.516

The total settlement is then:

$$S = 0.977(1.31)(2.85\ tsf)\left(0.516\frac{in}{tsf}\right) = 1.88\ inches$$

Gravity retaining wall

A gravity retaining wall relies on the mass of the wall and its geometry to support the earth pressure without tilting over or sliding. The weight of a gravity wall acts at the centroid of the wall, and the lateral earth pressure acts at one-third of the total wall height above the gravity wall base (for a perpendicular active-side face). The passive earth pressure acting on the non-active face of a gravity retaining wall is typically considered negligible. Concrete gravity retaining walls are considered appropriate for applications not requiring a retained height above approximately 10 feet.

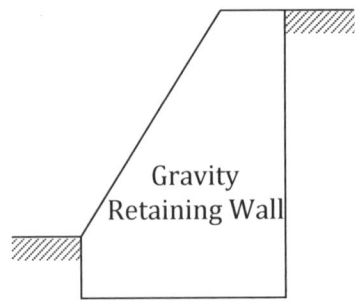

Problem

A 9-ft tall gravity retaining wall is 1 foot wide at the top and 5 feet wide at the base, as shown. Calculate the weight of the wall and the active pressure forces. Assume K_a is 1 and unit weight of soil is 100 lb/ft³.

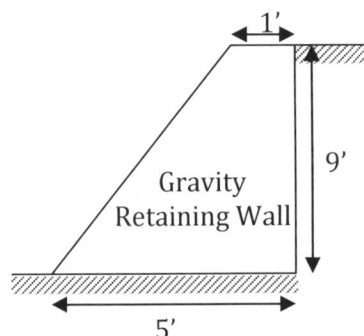

Solution
The weight of the wall is calculated by dividing the gravity wall into a rectangle and triangle to calculate the surface area, then multiplying the total surface area by 150 lb/ft³, the specific weight typically used for concrete:

$$Weight = \frac{150\ lb}{ft^3}\left((9ft \times 1ft) + \frac{1}{2}(9ft \times 5ft)\right)$$

$$Weight = 4725\ lb/foot$$

The centroid of the wall is calculated by determining the centroid of each area and taking the weighted average. Here the mean is calculated from the passive end of the wall:

$$\bar{x} = \frac{\Sigma CA}{\Sigma A} = \frac{2.67'\left[\frac{1}{2}(4')(9')\right] + 5.5'[(1')(9')]}{\left[\frac{1}{2}(4')(9')\right] + [(1')(9')]} = 3.61'$$

For granular backfill, the total active resultant, Ra, acting on the wall from the soil is at a distance of 3' from wall bottom is:

$$R_a = \frac{1}{2}K_a\gamma H^2 = \frac{100 lb/ft^3}{2} \times (9ft)^2 = 4050\ lb/ft$$

Cantilever retaining wall

A cantilever retaining wall has several components: the vertical stem, the horizontal base, and a key beneath the stem. The key is an optional component; it provides additional strength, but it is relatively time-consuming and costly feature for this type of wall. A cantilever retaining wall relies on the weight of the soil over the base heel and the soil resistance pressure beneath the base to prevent the wall from tilting over. The wall stem may also be tapered in thickness.

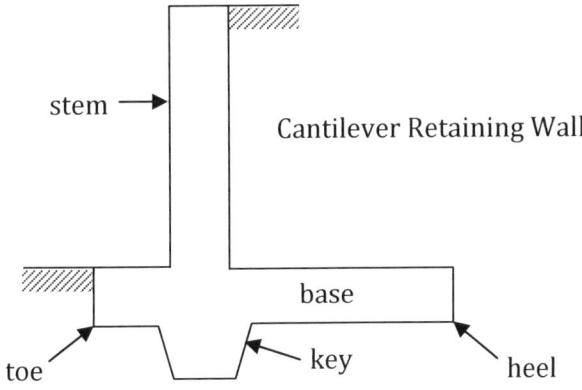

Problem

A cantilever retaining wall for clean sand and gravel backfill (unit weight of 105 lb/ft³) is shown below. Calculate the contributing moment about the toe from the weights of the concrete and soil sections. Assume a unit weight of concrete of 150 lb/ft³.

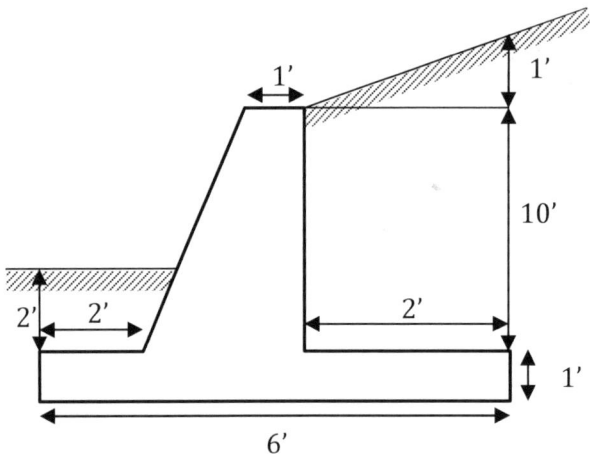

Solution

First, divide the soil and concrete into sections, then calculate the area, the weight (γ × Area), the centroid distance (x) from the toe, then the moment (W × x) for each section.

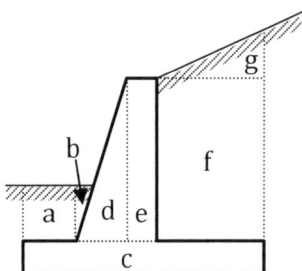

ID	Area, ft²	Weight, lb/ft	x, from toe, ft	M, lb*ft/ft
a	4.0	420	1.0	420
b	0.2	21	2.07	43
c	6.0	900	3.0	2700
d	5.0	750	2.67	2003
e	10.0	1500	3.5	5250
f	20.0	2100	5.0	10,500
g	1.0	105	5.33	560

The summation of the weights of is 5796 lb/linear foot of wall. The summation of the moment arms about the toe due to the weight from the pertinent concrete and soil sections, ∑Wx, for the cantilever retaining wall system shown below is 21,476 lb*ft/linear foot of wall. Determine the total moment about the toe, neglecting passive earth pressure.

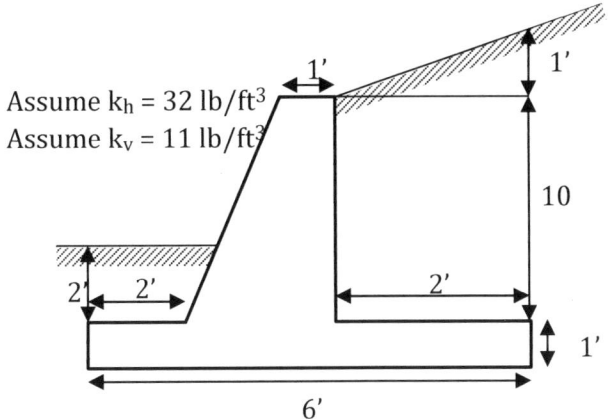

The horizontal earth pressure, $R_{a,h}$, and the vertical earth pressure, $R_{a,v}$, can be calculated from:

$$R_{a,v} = \frac{1}{2}k_v H^2 = \frac{1}{2}\left(11\frac{lb}{ft^3}\right)(12\ ft)^2 = 792\ lb/ft$$

$$R_{a,h} = \frac{1}{2}k_h H^2 = \frac{1}{2}\left(32\frac{lb}{ft^3}\right)(12\ ft)^2 = 2304\ lb/ft$$

$R_{a,h}$ acts at one-third of the total height (12 feet) above the retaining wall heel; thus, $R_{a,h}$ acts at the point 4 feet above the heel and toe. $R_{a,v}$ acts at the heel in a vertical direction, 6 feet away from the toe. The total moment about the toe is then:

$$M_{toe} = \Sigma Wx + (R_{a,v} * x_a) - (R_{a,h} * y_a)$$

$$M_{toe} = 21{,}476\ lb \times \frac{ft}{ft} + \left(\frac{792\ lb}{ft} \times 6ft\right) - \left(\frac{2304\ lb}{ft} \times 4ft\right)$$

$$M_{toe} = 21{,}476\ lb \times \frac{ft}{ft} + \left(\frac{792\ lb}{ft} \times 6ft\right) - \left(\frac{2304\ lb}{ft} \times 4ft\right)$$

$$M_{toe} = 17{,}012\ lb \times ft\ per\ linear\ foot\ of\ wall$$

Problem

A long, braced excavation in clay is 6 feet wide. Calculate the depth at which a heaving failure is likely to occur if the height of the excavation cut is 4 feet. Also, calculate the factor of safety of the slope against sliding. Assume the unit weight of the soil is 120 lb/ft³ and the cohesion is 300 lb/ft².

Solution
For conditions in which the height of an excavation cut (from the base of the cut to the top of the cut) is less than the excavation width, the equation to calculate the depth at which a heaving failure is likely to occur, H_c, in clay is shown below:

$$H_c = \frac{5.7 * c}{\gamma - \sqrt{2}\left(\frac{c}{B}\right)}$$

Where c is cohesion, B is the excavation width, and γ is the unit weight of the soil.

$$H_c = \frac{5.7 \times 300 \; lb/ft^2}{120 \; lb/ft^3 - \sqrt{2}\left(\frac{300 \; lb/ft^2}{6 \; ft}\right)} = 34.7 \; ft$$

The factor of safety of the slope against sliding is:
$$F_s = \frac{H_c}{H} = \frac{34.7 \; ft}{4 \; ft} = 8.7, which\; is\; acceptable$$

Problem

A long, braced excavation in clay is 6 feet wide. Calculate the depth at which a heaving failure is likely to occur if the height of the excavation cut is 12 feet. A surface surcharge loading of 100 lb/ft² exists. Calculate the factor of safety if the bracing stops at the base cut. Assume the unit weight of the soil is 120 lb/ft³ and the cohesion is 300 lb/ft².

Solution
For conditions in which the height of an excavation cut, H, (from the base of the cut to the top of the cut) is greater than the excavation width, the equation to calculate the depth at which a heaving failure is likely to occur, H_c, in clay is shown below:

$$H_c = \frac{N_c * c}{\gamma}$$

Where c is cohesion, γ is the unit weight of the soil, and N_c is the Skempton coefficient based on the H/B ratio. For an H/B ratio of 12 ft/6 ft = 2, the N_c coefficient is equal to 7.0 (tabulated value).

$$H_c = \frac{7.0 \times 300 \; lb/ft^2}{120 \; lb/ft^3} = 17.5 \; ft$$

The factor of safety, F, is calculated from the following, where q is the surface surcharge loading:

$$F = \frac{N_c \times c}{\gamma H + q} = \frac{7.0 \times 300 \; lb/ft^2}{120 \; lb/ft^3 (12 \; ft) + 100 \; lb/ft^2} = 1.36$$

Since the factor of safety is greater than 1.25, the bracing will not need to extend below the base of the cut.

Problem

A braced excavation in sand is 10 feet wide. Calculate the factor of safety if a) the water table is 15 feet below the excavation cut base and the unit weight of the drained sand immediately next to and below the bracing is identical, and b) the water table is steady at the excavation cut base elevation. Assumptions:
Drained unit weight of the soil = 95 lb/ft³
Internal angle of friction = 25°
Submerged unit weight of the soil = 115 lb/ft³
Bearing capacity factor, N_γ = 11
Active earth pressure coefficient = 0.30

Solution

For braced excavations in sand, the factor of safety is dependent on the unit weight of the soil, the internal angle of friction, and the water table location. If the water table is located at least the excavation width distance, B, below the excavation cut base, the following equation can be used to calculate the factor of safety, F:

$$F = 2N_\gamma k_a \tan\phi$$

Where N_γ is the bearing capacity factor, k_a is the active earth pressure coefficient, and ϕ is the internal angle of friction.

$$F = 2 * 11(0.30)\tan 25 = 3.08$$

For conditions where the water table is steady at the excavation cut base, the factor of safety, F, is increased by the multiplication of the ratio of submerged to drained soil unit weights:

$$F = 2N_\gamma \left(\frac{\gamma_{submerged}}{\gamma_{drained}}\right) k_a \tan\phi$$

$$F = 2 * 11\left(\frac{115\ lb/ft^3}{95\ lb/ft^3}\right) 0.30 \times \tan 25 = 3.73$$

Both of the factors of safety calculated are acceptable.

Problem

Sketch the design pressure envelope for a braced excavation cut in sand. Also give the equations for the maximum lateral pressure, p_{max}, and the coefficient of active earth pressure, k_a.

<u>Solution</u>
Two design pressure envelopes for a braced excavation cut in sand are shown below. The Peck design pressure envelope is the more common for braced cuts in sand. Below, H is the height of the cut from the base to the top, γ is the drained unit weight of sand, and ϕ is the drained internal angle of friction.

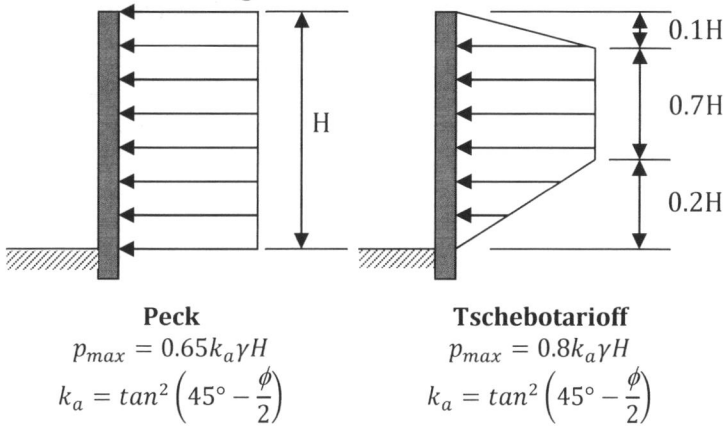

Peck
$p_{max} = 0.65 k_a \gamma H$
$k_a = \tan^2\left(45° - \frac{\phi}{2}\right)$

Tschebotarioff
$p_{max} = 0.8 k_a \gamma H$
$k_a = \tan^2\left(45° - \frac{\phi}{2}\right)$

Problem

Sketch the design pressure envelope for a braced excavation cut in soft to medium clay. Also give the equations for the maximum lateral pressure and the coefficient of active earth pressure.

Solution

The Peck design pressure envelope for a braced excavation cut in soft to medium clay ($\gamma H/c > 4$) is shown. Below, p_{max} is the maximum lateral pressure, k_a is the coefficient of active earth pressure, H is the height of the cut from the base to the top, γ is the drained unit weight of sand, and c is cohesion.

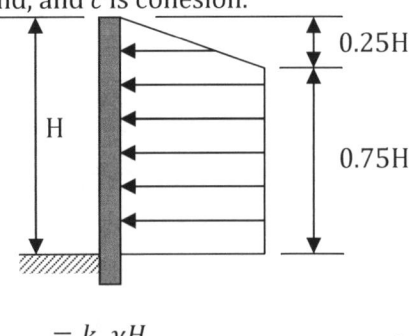

$$p_{max} = k_a \gamma H$$
or
$$0.2\gamma H \leq p_{max} \leq 0.4\gamma H,$$
whichever is greater

$$k_a = 1 - \frac{4c}{\gamma H}$$

Problem

Sketch the design pressure envelope for a braced excavation cut in stiff clay. Also give the equations for the maximum lateral pressure and the coefficient of active earth pressure.

Solution

The Peck design pressure envelope for a braced excavation cut in stiff clay ($\gamma H/c \leq 4$) is shown. Below, p_{max} is the maximum lateral pressure, k_a is the coefficient of active earth pressure, H is the height of the cut from the base to the top, γ is the drained unit weight of sand, and c is cohesion.

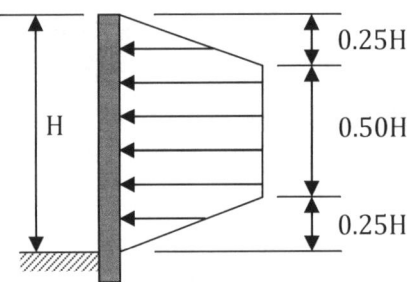

$$0.2\gamma H \leq p_{max} \leq 0.4\gamma H$$

$$k_a = 1 - \frac{4c}{\gamma H}$$

Structural

Loads

Live loads refer to forces that are variable during the structure's lifetime. Live loads are not permanent. Examples of live loads include forces attributed to people, vehicles, etc. Environmental loads such as snow and wind loadings can be categorized as live loads, although environmental loads may be considered separately.

Dead loads refer to loads within or on a structure that are permanent. Dead loads include the weights of the materials used in construction. Examples of dead loads include the weights of the floors, walls, roof, and any permanent equipment used within the structure.

Construction loads are forces that are present during the construction phase. Some construction loads may be incorporated permanently, but the majority of construction loads are present only during the construction phase of the project. Construction loads may be substantially greater than the expected permanent loads and thus must be considered during design. Examples of construction loads include construction equipment and activities such as mobilization and placement of equipment.

Problem

Calculate the resultant forces for the simply supported beam in the example below.

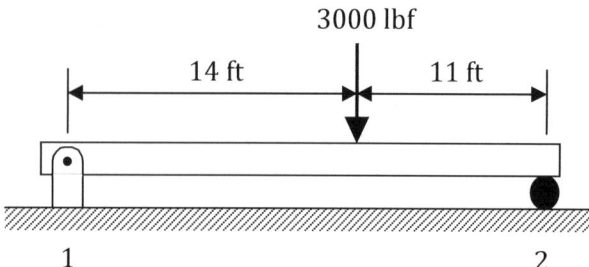

Solution
R_1 is a pinned support and has two components. R_2 is a roller support and has one component. The free-body diagram is:

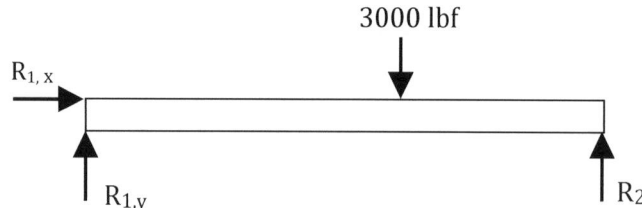

Taking Moment about Point 1:
$\sum M_1 = (3000 \text{ lbf})(14 \text{ ft}) - R_2(25 \text{ ft}) = 0$; **$R_2 = 1680$ lbf**
Equilibrium Equation for Vertical Forces:

$\Sigma F_y = R_{1,y} + R_2 - 3000 \text{ lbf} = 0$
   $R_{1,y} + 1680 \text{ lbf} - 3000 \text{ lbf} = 0$; **$R_{1,y}$ = 1320 lbf**
Equilibrium Equation for Horizontal Forces:
$\Sigma F_x = R_{1,x} + 0 = 0$; **$R_{1,x}$ = 0** (no horizontal forces)

Problem

Determine the equivalent force (magnitude and location) of the distributed load on the beam below.

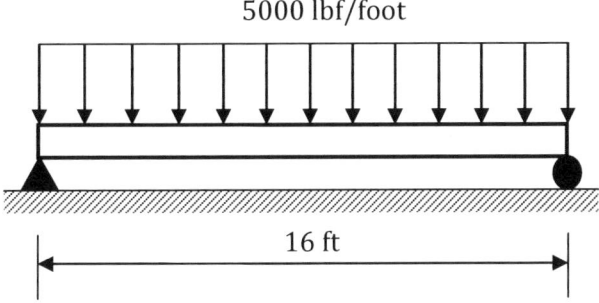

Solution
The equivalent force of a distributed load acts at the centroid of the distributed load area. The centroid of a rectangular distributed load is located at a distance of L/2, where L is the length of the distributed load along the beam. For this problem, the equivalent force acts at a distance of 8 feet from either end of the beam (16 feet / 2). Here, the magnitude of the equivalent force, F_e, is calculated by multiplying the load per foot by the total length of the load:

$$F_e = \left(5000\frac{lbf}{ft}\right)(16\ ft) = 80{,}000\ lbf$$

The equivalent force is thus:

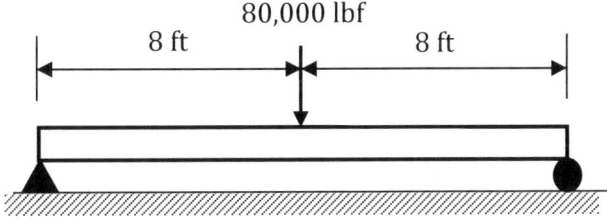

Problem

Determine the equivalent force (magnitude and location) of the distributed load on the beam below.

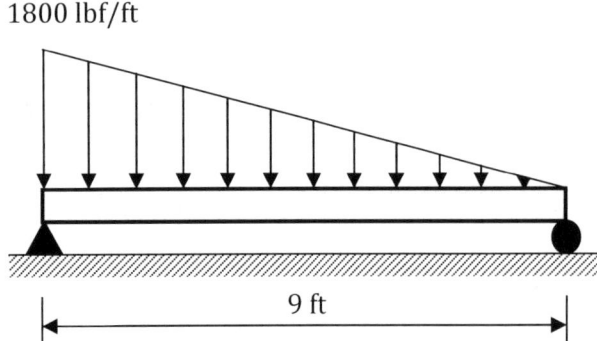

Solution

The equivalent force of a distributed load acts at the centroid of the distributed load area. The centroid of a triangular distributed load is located at a distance of 2L/3 from the smallest end (where the load is equal to 0), where L is the length of the distributed load along the beam. For this problem, the equivalent force acts at a distance of (2 × 9 ft)/3 = 6 feet from the right side of the beam. The magnitude of the equivalent force, F_e, is the area of the triangular load:

$$F_e = \frac{1}{2}\left(1800\frac{lbf}{ft}\right)(9\ ft) = 8{,}100\ lbf$$

Problem

Determine the two equivalent forces (magnitude and location) of the distributed load between the supports of the beam below.

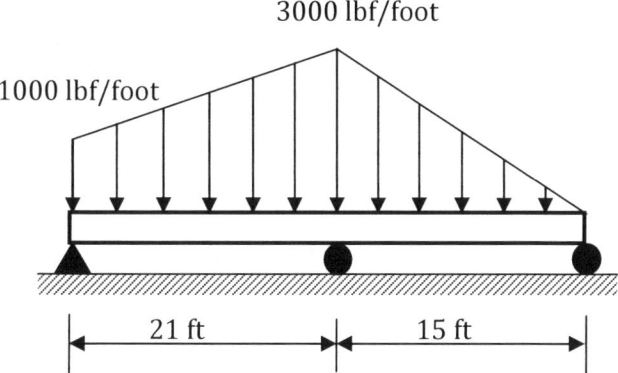

Solution

For the trapezoidal distributed load, the centroid is located at (h/3)[(b+2t)/(b+t)], where h is the length of the distributed load along the beam, b is the longer base, and t is the shorter base: (21 ft/3)[(3000 ft + 2 × 1000 ft)/(3000 ft + 1000 ft)] = 8.75 ft from the longer base. For the triangular distributed load, the centroid is located (2 × 15ft)/3 = 10 feet from the right side. The area of the trapezoidal load is (1000 lbf/ft × 21 ft) + (1/2)(2000 lbf/ft)(21 ft) = 42,000 lbf. The area of the triangular load is (1/2)(3000 lbf/ft)(15 ft)=22,500 lbf. The equivalent forces on the beam are thus:

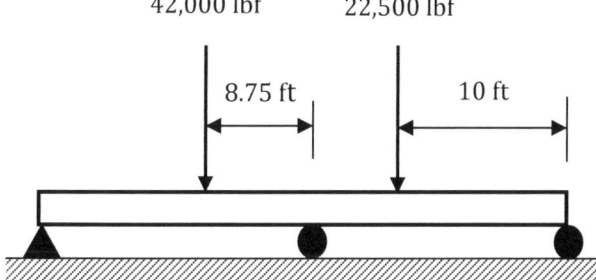

Problem

Calculate the total dead load (per foot) on a beam that supports the following: a 5-inch thick stone concrete slab that is 8 feet wide and an 8-inch clay brick wall that is 9 feet tall. Acoustical fiberboard covers the bottom side of the stone concrete slab. Assume the slab is centered on the beam, and the clay brick wall is centered along the length of the beam.

Solution

Per ASCE 7-05, the minimum design dead loads for the materials are: stone concrete slab = 12 lbf/ft²*in, 8-in clay brick wall = 79 lbf/ ft², and acoustical fiberboard = 1 lbf/ ft².
The dead load of the stone concrete slab along the beam is:

$$12 \frac{lbf}{ft^2 \times in} (8\,ft)(5\,in) = 480\,lbf/ft$$

The dead load of the 8-inch clay brick wall along the beam is:

$$79\frac{lbf}{ft^2}(9\,ft) = 711\,lbf/ft$$

The dead load of the acoustical fiberboard along the beam is:

$$1\frac{lbf}{ft^2}(8\,ft) = 8\,lbf/ft$$

The total dead load on the beam is:

$$480\frac{lbf}{ft} + 711\frac{lbf}{ft} + 8\frac{lbf}{ft} = 1199\,lbf\,per\,ft\,of\,beam$$

Problem

ASCE 7-05 allows for reduction in live loads with some exceptions. List a few of those exceptions.

Solution
Minimum uniformly distributed live loads may be reduced as long as:
- the live uniform load is not a roof load
- the live uniform load is 100 psf or less
- the live uniform load is not in a passenger car garage
- the live uniform load is not in a public assembly area, such as a library, theater, or school classroom
- the reduced design live load is equal to or greater than 50% of the unreduced design live load for members supporting one floor
- the reduced design live load is equal to or greater than 40% of the unreduced design live load for members supporting two or more floors

An exception to the exceptions above is that live loads for members supporting two or more floors may be reduced by 20%, even if the live loads exceed 100 psf and even if the live load is present in a passenger car garage.

Problem

Determine if the structure below is statically determinate or statically indeterminate.

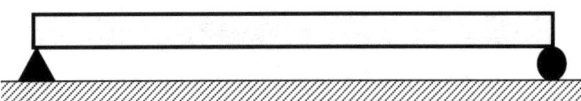

Solution
The forces acting on the beam from the supports are shown below:

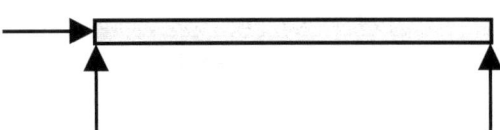

There are three forces acting on the beam from the supports. The beam is a solid beam and consists of one component. A structure is statically determinate if the number of forces acting on the beam equals three times the number of components:
$No.\,of\,Forces = 3 * No.\,of\,structure\,components$
For this structure: 3 = 3 × 1, so the structure is statically determinate.

This indicates that the forces acting on the beam from the supports can be determined using equilibrium equations.

Problem

Determine if the structure below is statically determinate or statically indeterminate.

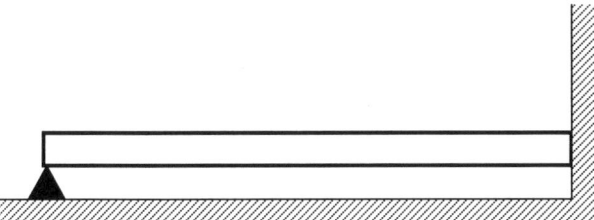

Solution
The forces acting on the beam from the support and the attachment to the wall are shown below:

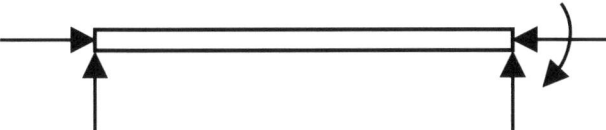

There are five forces acting on the beam from the support and the wall attachment. The beam is a solid beam and consists of one component. A structure is statically indeterminate if the number of forces acting on the beam is greater than three times the number of components:

No. of Forces $> 3 *$ No. of structure components

For this structure: $5 > 3 \times 1$, so the structure is statically indeterminate. This indicates that the forces acting on the beam from the supports cannot be determined using equilibrium equations. Two additional equations would be needed to solve for the forces.

Problem

Determine if the structure below is stable or unstable.

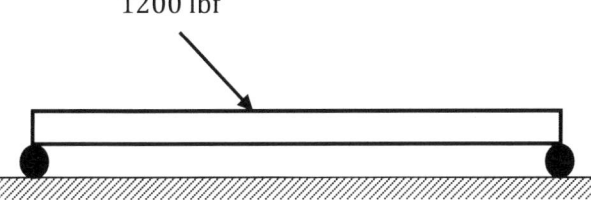

Solution
A structure's stability is an indication of how well the structure or its member(s) are constrained by their supports. Stability is necessary for structure equilibrium. Structure instability is either due to partial or improper constraints. The forces from the supports of the beam below are vertical only. Since the 1200 lb force being applied on the beam is at an angle, it has both a vertical and horizontal component, and the summation of the forces in

the horizontal cannot equal zero. Since the horizontal component cannot be supported by the beam supports, the structure is not in equilibrium, and thus is unstable.

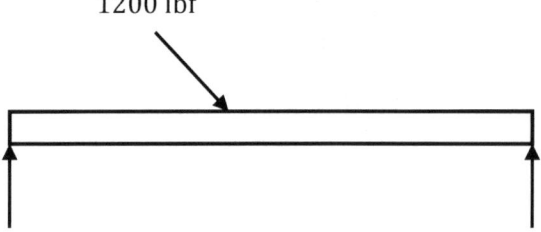

Problem

Determine the vertical and horizontal forces at the supports B and D using the joints method.

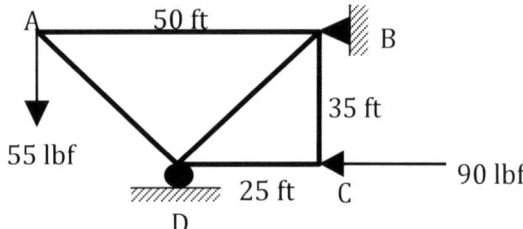

Solution
The reactions at B and D are assumed and shown below:

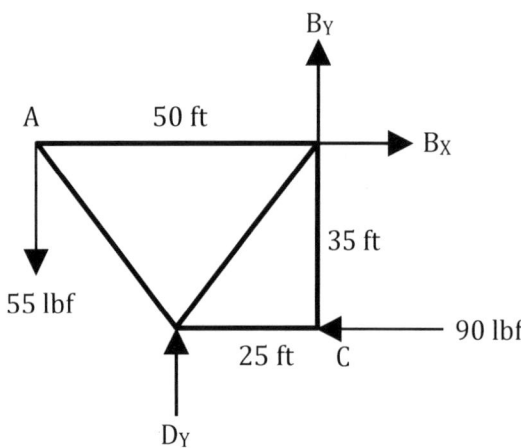

$\sum F_X=0$: $B_X - 90$ lbf $= 0$; $B_X = 90$ lbf
$\sum M_B=0$: $(90 \text{ lbf} \times 35 \text{ ft}) + (D_Y \times 25 \text{ ft}) - (55 \text{ lbf} \times 50 \text{ ft}) = 0$; $D_Y = -16$ lbf
$\sum F_Y=0$: $B_Y + D_Y - 55$ lbf $= 0$; $B_Y - 16$ lbf $- 55$ lbf $= 0$; $B_Y = 71$ lbf

Problem

Determine the forces in members AB and AD. Also identify whether the members are in tension or compression.

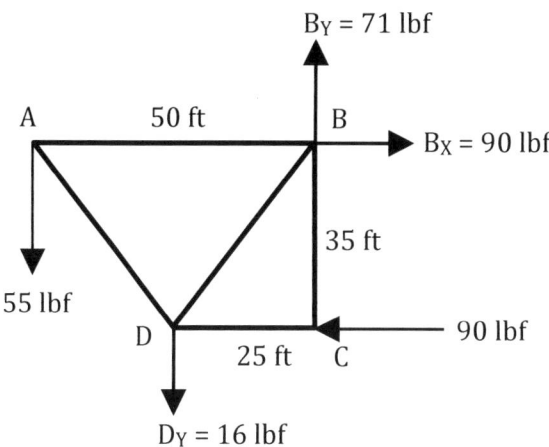

Solution
First, determine the interior angle A between members AB and AD:
$$\tan A = \frac{35°}{25°} = 54.46°$$
The forces of members AB and AD can then be calculated from the equilibrium equations, assuming the directions of the forces in the respective members.

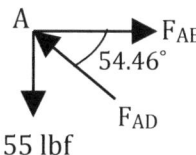

$\Sigma F_Y=0$: -55 lbf $+ F_{AD} \sin(54.46°) = 0$; $F_{AD} = 67.59$ lbf (Compression)
$\Sigma F_X=0$: $F_{AB} - F_{AD} \cos(54.46°) = 0$; $F_{AB} - (67.59$ lbf$) \cos(54.46°) = 0$
$F_{AB} = 39.29$ lbf (Tension)

Problem

Calculate the resultant force in member AB and the x- and y-components of the force.

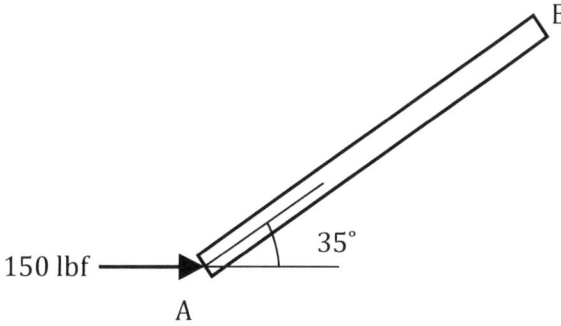

Solution
The x-component of the resultant force, AB_X is 150 lbf, since this force is applied in the horizontal direction.

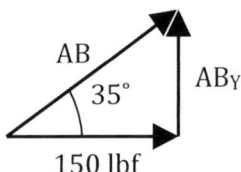

The y-component of the resultant force is calculated from:
$\tan 35 = \dfrac{AB_Y}{150\ lbf}; AB_Y = 105.0\ lbf$

The resultant force AB can be calculated from:
$\cos 35 = \dfrac{150\ lbf}{AB}; AB = 183\ lbf$

Alternatively, AB could be calculated from:
$AB = \sqrt{AB_X^2 + AB_Y^2} = \sqrt{(150\ lbf)^2 + (105\ lbf)^2} = 183\ lbf$

Problem

For the truss below, determine by inspection which members, if any, are zero force members.

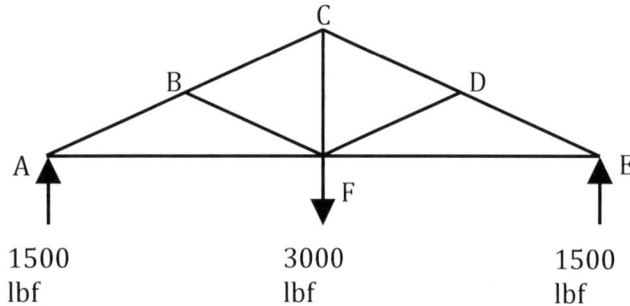

Solution
Zero force members occur in trusses when any of the following occur:
1) For joints consisting of two members, both members are zero force members when no support or external force is applied at the joint.
2) For joints consisting of three members with two of those members being collinear, the third non-collinear member is a zero force member when no support or external force is applied at the joint.
3) For joints consisting of two members when one of the members is situated along the line of action caused by the loading, the other member is a zero force member.

For the truss above, the joint at B is an unloaded joint consisting of three members, with members AB and BC being collinear. Thus member BF is a zero force member. Similarly, DF is a zero force member since joint D is unloaded and members CD and DE are collinear.

Problem

Determine the forces in members AB, AF, DE, and EF in the truss below. Determine if the members are in compression or tension.

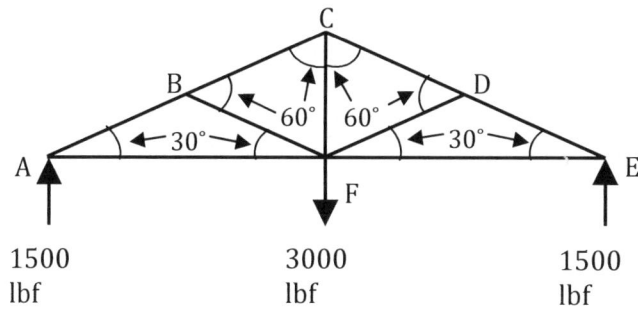

Solution

The forces in members AB and AF can be calculated by examining Joint A:

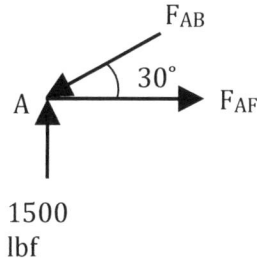

$\sum F_Y=0$: 1500 lbf - $F_{AB} \sin(30°) = 0$; F_{AB} = 3000 lbf. Since the force F_{AB} is pushing on Joint A, the member AB is in compression.

$\sum F_X=0$: $F_{AF} - F_{AB} \cos(30°) = 0$; F_{AF} – (3000 lbf) × cos(30°) = 0; F_{AF} = 2598.08 lbf. Since the force F_{AF} is pulling away from Joint A, the member AF is in tension. By inspection, the truss is symmetrical since the angles and forces on each side of the truss are mirrors of each other. Thus, the forces in the members will be identical to their counterpart: F_{DE} = 3000 lbf (C) and F_{EF} = 2598.08 lbf (T).

Problem

Determine the forces in members BC and CF in the truss below, given that the force in member AB is 3000 lbf (compression). Determine if the members are in compression or tension.

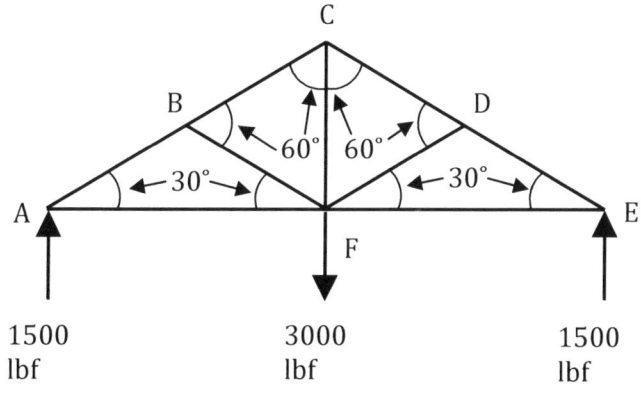

Solution

The forces in the members BC and CF can both be determined by inspection. At Joint B, three members are present. Since members AB and BC are collinear, and no load is being supplied or supported at joint B, the non-collinear member BF is a zero force member. Thus the force of member AB is equal to the force of member BC, 3000 lbf. Like member AB, member BC is in compression. Since the truss is symmetric, the forces on the right side will be identical to those on the left side of the truss. Member DF then is also a zero-member force. Because both members BF and DF are zero force members, the member CF must equal the force shown at Joint F, 3000 lbf. The member CF will be in tension (force is pulling away from joint).

Shear diagram

A shear diagram is a diagram that illustrates the shear at any given point along a beam subjected to a loading. A shear diagram is used to determine the maximum and minimum shear forces which are of particular importance when designing beams. The shear forces are plotted along the y-axis, and the beam distance is along the x-axis of the diagram. When constructing a shear diagram, the calculated shear at any given point is equal to the sum of forces from the far left end of the beam to the referenced point and is also equal to the slope of the moment diagram line at that location ($V = dM/dx$). Loads and reactions that are acting in an upward direction are positive forces. At points of concentrated loads, the shear is indeterminate and is shown as a vertical line on the shear diagram. Between concentrated loads, the shear diagram will be horizontal. For uniformly distributed loads, the shear diagram will show a constant slope over the region under a uniform load.

Bending moment diagram

A bending moment diagram illustrates the bending moment at any given point along a beam subjected to a loading. The maximum and minimum bending moment forces are of particular importance when designing beams (maximum bending moment is at the point where shear equals zero). The bending moment forces are plotted along the y-axis, and the beam distance is along the x-axis of the diagram. When constructing a moment diagram, the calculated bending moment at any given point is equal to the sum of moment and couple forces from the far left end of the beam to the referenced point and is also equal to the area under the shear diagram line up to the referenced point ($M = \int V dx$). Moment and couple forces that are acting in a clockwise direction are positive forces. Between concentrated loads, the moment diagram will show a constant slope over that region. For uniformly distributed loads, the moment diagram will show a parabolic upward curve over the region under a uniform load.

Problem

Sketch the shear diagram for the concentrated load F on the concrete beam as shown below.

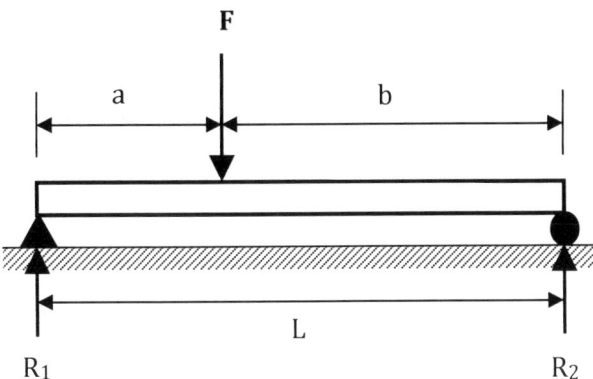

Solution

To sketch the shear diagram, the values of shear must first be determined. From the equilibrium equations, taking the moment separately about R_1 and R_2 yields the two following equations:

$$R_1 = \frac{F * b}{L} \text{ and } R_2 = \frac{F * a}{L}$$

Cutting the beam to the left of F and looking at the free body diagram, it can be seen that the shear on the left side is equal to R_1. Cutting the beam to the right of F and looking at the free body diagram, the shear to the right is equal to $R_1 - F$, which is equal to $-(F \times a)/L$. The shear diagram can then be drawn:

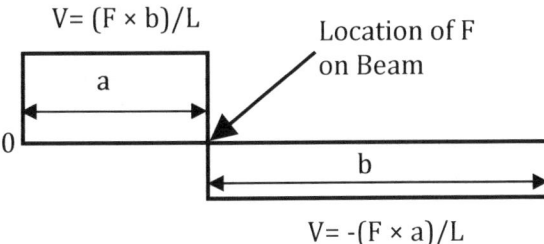

Problem

Sketch the shear diagram for the uniformly distributed load f on the concrete beam as shown below.

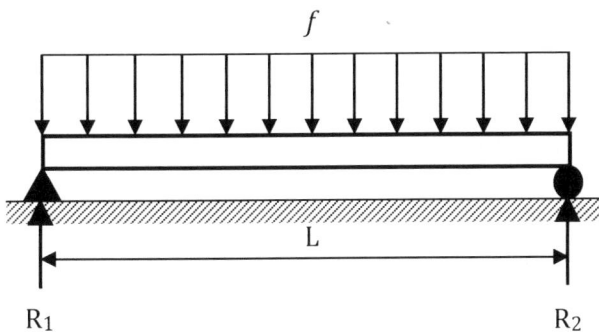

Solution

From the equilibrium equations, the forces R_1 and R_2 are determined to be equal to $(f \times L/2)$. The shear diagram of a uniformly distributed load will be a sloped line (with constant slope). The endpoints of the sloped line will equal the shear values at the far left of the beam ($x = 0$) and at the far right of the beam ($x = L$). The shear at a point x away from R_1 is equal to $R_1 - (f \times x)$, which is equal to $(f \times L/2) - (f \times x)$. Substituting 0 then L in for x, the shear values at the far left and far right are $(f \times L/2)$ and $-(f \times L/2)$, respectively. The shear diagram can then be drawn:

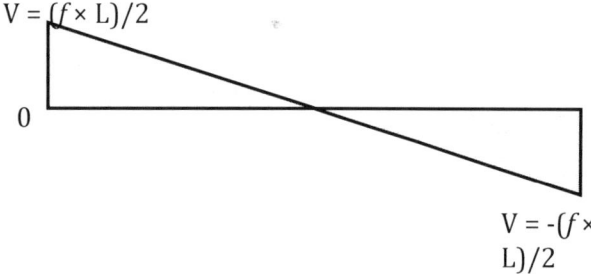

Problem

Sketch the shear diagram for the concentrated loads on the concrete beam as shown below. Assume F_1 is less than R_1.

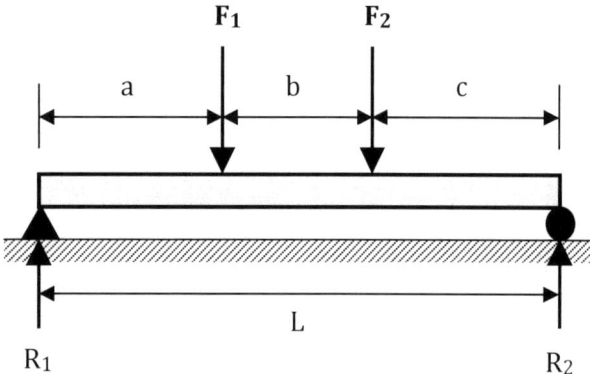

Solution

Looking at the free-body diagrams at each of the three segments in between the supports and forces, the following shear values can be determined: for first segment (between R_1 and F_1): $V = R_1$, for second segment (between F_1 and F_2): $V = R_1 - F_1$, and for the third segment, looking at the free-body diagram from the right-hand side of the beam, $V = -R_2$. Assuming F_1 is less than R_1, the shear diagram is shown below. It is important to note that the shear for the third segment (between F_1 and F_2) is also equal to $R_1 - F_1 - F_2$.

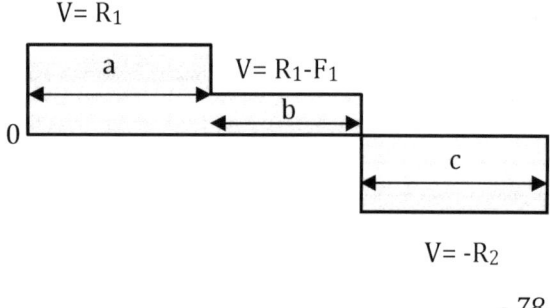

Problem

Sketch the moment diagram for the concrete beam and the corresponding shear diagram shown below.

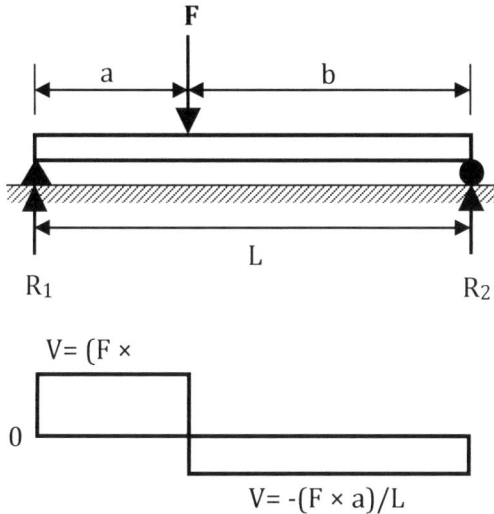

Solution

Looking at the free body diagrams to the left of force F (at a distance x from the left end), the moment of the left-hand side of the beam can be determined to be $R_1 \times x$. From the shear diagram, we can see that $R_1 = (F \times b)/L$, so the moment on the left-hand side is equal to $(F \times b \times x)/L$. Looking at the free body diagram that results from cutting the beam to the right of force F (a distance x from the left end), the moment on the right-hand side of the beam is equal to $R_1 \times x - F(x-a)$, which can be rewritten as $(F \times a/L)(L-x)$. The maximum moment is equal to $(F \times a \times b)/L$ and occurs at the location the concentrated load F is applied. The moment diagram is then:

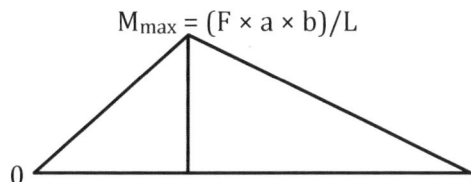

Problem

Sketch the moment diagram for the concrete beam and the corresponding shear diagram shown below.

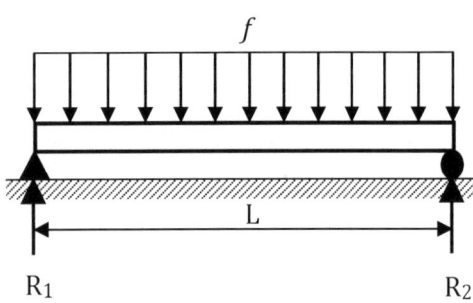

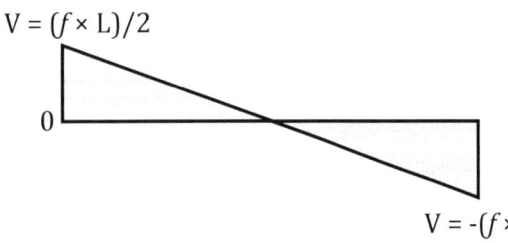

Solution

Cutting the beam at a distance x from the left end of the beam and looking at the free-body diagram created at the left end, the moment is equal to $R_1 \times x - (f \times x(x/2))$. Since the load is uniformly distributed across the entire length of the beam, the moment at the opposite end will be equal to the negative moment value of the left end of the beam. The moment diagram will consist of a parabolic curve (mirrored about the midpoint of the beam) with a maximum moment at the midpoint equal to the area of the shear diagram from the end of the beam (x = 0) to the beam midpoint (x = L/2). This area is triangle and is equal to (1/2)(base)(height): $(1/2)(L/2)(f \times L/2) = f \times L^2/8$. The moment diagram is then:

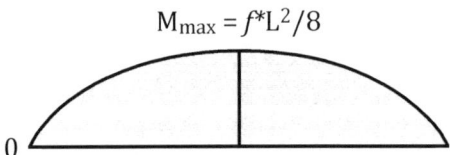

Problem

Sketch the moment diagram for the concrete beam and the corresponding shear diagram shown below.

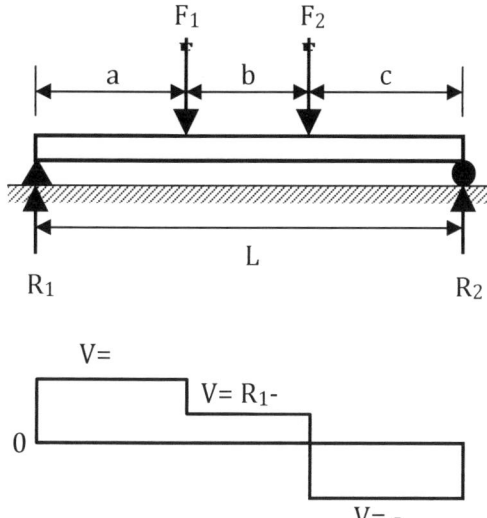

Solution

The moment diagram for a beam with more than one concentrated loads will consist of constant sloped lines between the supports and the concentrated loads. The value of the moment at the endpoints of the sloped lines can be determined by taking the free body diagram at points (a distance x from the left-hand side of the beam) before and after each concentrated load. Between R_1 and F_1, the moment is $R_1 \times x$, and M_1 is thus $R_1 \times a$. Between F_1 and F_2, the moment is $(R_1 \times x) - F_1(x-a)$. Between F_2 and R_2, the moment is $R_2(L-x)$, and M_2 is thus $R_2 \times c$. The maximum moment occurs at the location of the concentrated load at which the shear goes from positive to negative (in this case, at F_2). The moment diagram is then:

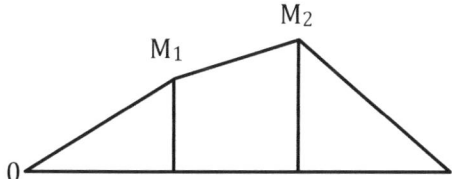

Flexural stress (bending stress) of a concrete beam

Flexural stress (bending stress) is the normal stress a beam experiences under a transverse force (a force acting at a right angle to the beam). The transverse force causes the beam surface furthest away from the force to experience tensile stress (beam lengthening). The surface of the beam closest to the force will experience compressive stress (beam shortening). The neutral axis is a horizontal plane of zero normal stress that passes through the centroid of the beam cross-section. The flexural stress increases with increased distance from the neutral axis.

Flexural stress in a supported beam:

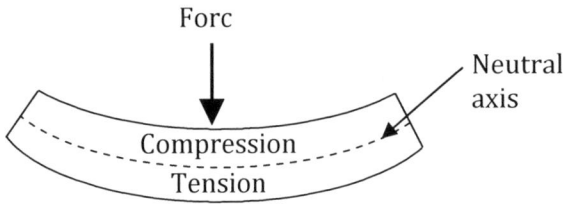

Problem

*From the shear and moment diagrams for a loaded concrete beam, the maximum shear and maximum bending moment are determined to be 512 lbf and 1275 lbf*ft, respectively. Determine the maximum shear stress and the maximum bending stress if the beam has a circular cross section with a radius of 6 inches.*

Solution
The maximum shear stress, τ_{max} in a circular beam is calculated using the following equation:
$$\tau_{max} = \frac{4V}{3A}$$
Where V is the vertical shear at location in question and A is the beam cross-sectional area.
The maximum shear stress is then:
$$\tau_{max} = \frac{4 \times 512 \ lbf}{3 \times \pi(6 \ in)^2} = 6.0 \ lbf/in^2$$
The maximum bending stress is calculated from the following equation:
$$\sigma_{b,max} = \frac{M \times c}{I_c}$$
Where M is the bending moment at location in question, c is the distance from the neutral axis to the extreme surface, and I_c is the beam's cross section centroidal moment of inertia. I_c for a circular beam is $\pi d^4/64 = \pi(12 \ in)^4/64 = 1018 \ in^4$. The maximum bending stress is then:
$$\sigma_{b,max} = \frac{1275 lbf \times ft \times 6 \ in \times (12in/1ft)}{1018 \ in^4} = 90.2 \ lbf/in^2$$

Problem

*From the shear and moment diagrams for a loaded concrete beam, the maximum shear and maximum bending moment are determined to be 450 lbf and 1145 lbf*ft, respectively. Determine the maximum shear stress and the maximum bending stress if the beam has a rectangular cross section as shown below:*

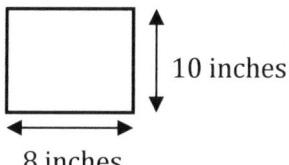

8 inches

Solution

The maximum shear stress, τ_{max} in a rectangular beam is calculated using the following equation:

$$\tau_{max} = \frac{3V}{2A}$$

Where V is the vertical shear at location in question and A is the beam cross-sectional area. The maximum shear stress is then:

$$\tau_{max} = \frac{3 \times 450 \, lbf}{2 \times 8in \times 10in} = 8.4 \, lbf/in^2$$

The maximum bending stress is calculated from the following equation:

$$\sigma_{b,max} = \frac{M \times c}{I_c}$$

Where M is the bending moment at location in question, c is the distance from the neutral axis to the extreme surface, and I_c is the beam's cross section centroidal moment of inertia. I_c for a rectangular beam is $bh^3/12$ = $((8 \text{ in}) \times (10in)^3)/12$ = 667 in⁴. The maximum bending stress is then:

$$\sigma_{b,max} = \frac{1145 lbf \times ft \times 5 \, in \times (12in/1ft)}{667 \, in^4} = 103.0 \, lbf/in^2$$

Shear capacity of a reinforced concrete beam

A reinforced beam's shear capacity (nominal beam strength), V_n, is the summation of the beam's concrete strength (nominal shear strength of normal-weight concrete), V_c, and the nominal shear strength of the shear reinforcement, V_s. A strength reduction factor is applied to V_n to obtain the factored applied shear (V_u). The simplified equation to calculate the beam's concrete strength is:

$$V_c = 2db_w\sqrt{f_c'}$$

Where d is the distance, in inches, from the centroid of tension reinforcement to the extreme compression fiber, b_w is the web width in inches, and f_c' is the proposed specified design strength in pounds per square inch. The nominal shear strength of the shear reinforcement is calculated from:

$$V_s = \frac{A_v f_y d}{s}$$

Where A_v is the area of shear reinforcement, f_y is the specified yield strength of the reinforcement, and s is the horizontal distance (spacing) between the bars.

Shear in W-shape steel beams

In W-Shape steel beams, the thickness of the web bears the shear in its entirety. The average shear stress in the steel beam web, f_v, is calculated from:

$$f_v = \frac{V}{A_w} = \frac{V}{dt_w}$$

Where V is the maximum vertical shear, A_w is the gross area of the web, d is the depth of the beam, and t_w is the web thickness. The maximum allowable shear stress in the steel beam web, F_v, is equal to $0.4F_y$, where F_y is the yield stress in the member. Shear stress determination in steel beams becomes an important factor for large loads within short spans or at supports.

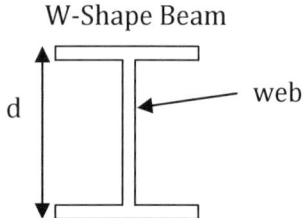

W-Shape Beam

Determining the maximum stirrup spacing in reinforced concrete

$For\ V_s \leq 4db_w\sqrt{f_c'}$ (note: this value equals $2 \times V_c$),
The stirrup spacing shall not exceed the lesser of either 24 inches or d/2.

$For\ V_s > 4db_w\sqrt{f_c'}$ (note: this value equals $2 \times V_c$),
The stirrup spacing shall not exceed the lesser of either 12 inches or d/4. The second equation used to check stirrup spacing, s, is:

$$s = \frac{A_v f_y d}{V_{s,req}}$$

Where the $V_{s,\,req}$ is equal to $(V_u/\phi) - V_c$. For these equations, V_s is the nominal shear strength of the shear reinforcement, d is the distance, in inches, from the centroid of tension reinforcement to the extreme compression fiber, b_w is the web width in inches, f_c' is the proposed specified design strength in pounds per square inch, A_v is the area of shear reinforcement, f_y is the specified yield strength of the reinforcement, V_u is the factored applied shear, and V_c is the beam's concrete strength. The stirrup spacing is the lesser of the two values calculated.

Average tensile stress in an axially loaded structural steel tension member

In an axially loaded structural steel tension member, the average tensile stress, f_t, is equal to P/A, where *P* is the applied load and *A* is the cross-sectional area, normal to the load. The cross-sectional area can be a gross area, a net area, or an effective net area value. See below for a pictorial explanation of gross and net areas. The effective net area is calculated by applying a reduction coefficient to either the gross area or net area, depending on the situation.

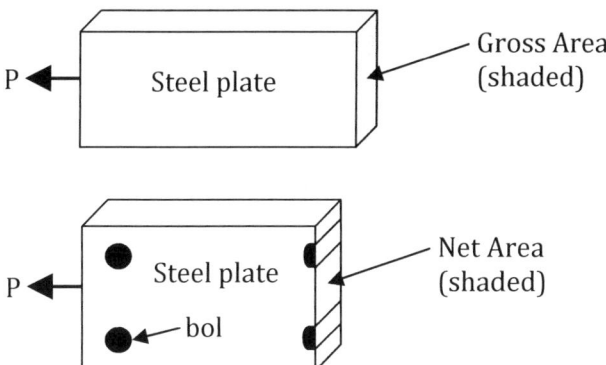

Problem

Two A36 steel plates are bolted together. The net areas of the potential failure paths from the bolt hole locations include the following: Path A of 6.75 square inches, Path B of 5.32 inches, and Path C of 6.24 inches. Determine the tensile capacity with respect to yielding and with respect to fracture if the steel yield strength is 36 ksi, the tensile strength is 58 ksi, the gross area of the plate is 7.5 square inches, and all of the connections are in the same plane. Also, determine the actual tensile capacity (neglecting block shear failure).

Solution
The tensile capacity with respect to yielding, $P_{t,yielding}$, is calculated from the following equation:
$P_{t,\ yielding} = 0.6 F_y A_g$
Where F_y is the steel yield strength, and A_g is the gross area of the steel plate. Thus,
$P_{t,\ yielding} = 0.6(36\ ksi)(7.5\ in^2) = 162\ kips$
The tensile capacity with respect to fracture, $P_{t,fracture}$, is calculated from the following equation:
$P_{t,\ fracture} = 0.5 F_u A_e = 0.5 F_u U A_n$
Where F_u is the tensile strength, U is a reduction coefficient, and A_n is the net area of the steel plate. For connections that are in the same plane, U is equal to one. The net area used is the least of the net areas determined from the potential failure paths. Thus,
$P_{t,\ fracture} = 0.5(58\ ksi)(1)(5.32\ in^2) = 154.28\ kips$
The actual tensile capacity (neglecting block shear failure) is the lesser of the two tensile capacities calculated. Thus, the actual tensile capacity is 154.28 kips (fracture).

Problem

Determine the allowable block shear strength in a steel assembly with the following characteristics:
Net shear area = 1.39 square inches
Net tension area = 0.28 square inches
Tensile strength = 60 kips
All of the connections are in the same plane.

Solution
The allowable rupture strength in shear is equal to $0.3 F_u A_v$, where F_u is the steel tensile strength and A_v is the net shear area. For the given information, this value is:

$0.3(60\ kips)(1.39\ in^2) = 25.02\ kips$

The allowable rupture strength in tension is equal to $0.5F_uUA_t$, where U is a reduction coefficient and A_t is the net tension area. Since all of the connections in the steel assembly are in the same plane, U is equal to one. For the given information, this value is:

$0.5(60\ kips)(1)(0.28\ in^2) = 8.4\ kips$

The allowable block shear strength, $P_{t,\ block\ shear}$ is calculated from the addition of the allowable rupture strengths in shear and tension:

$P_{t,\ block\ shear} = 0.3F_uA_v + 0.5F_uUA_t$

$$P_{t,\ block\ shear} = 25.02\ kips + 8.4\ kips = 33.42\ kips$$

Problem

A 2-inch diameter steel cylinder is subjected to a 350-kip axial tensile load. The length of the cylinder prior to the load is 8 feet. After the tensile load, the length increased by 0.36 inches. Calculate the modulus of elasticity of the steel, assuming the material behaves linearly elastically.

Solution

The first step is to calculate the longitudinal stress, σ, which is equal to P/A, where P is the axial force on the cylinder, and A is the cross-sectional area of the cylinder. The longitudinal stress is thus:

$$\sigma = \frac{350\ kips}{\pi \left(\frac{2\ in}{2}\right)^2} = 111.41\ ksi$$

Next, the axial strain, ε, is calculated from:

$$\epsilon = \frac{\delta}{L}$$

Where δ is the total elongation, and L is the original length of the cylinder.

$$\epsilon = \frac{0.36\ inches}{8\ ft\ \left(\frac{12\ in}{1\ ft}\right)} = 3.75 \times 10^{-3}$$

The modulus of elasticity, E, may then be calculated from:

$$E = \frac{\sigma}{\epsilon} = \frac{111.41\ ksi}{3.75 \times 10^{-3}} = 2.97 \times 10^4\ ksi$$

Problem

A 2-inch diameter steel cylinder is subjected to an axial tensile load. The length of the cylinder prior to the load is 8 feet. After the tensile load, the length increased by 0.36 inches (0.03 ft) and the diameter of the cylinder decreased by 0.002 inches. Calculate Poisson's ratio for the cylinder material. The calculated axial strain is 3.75×10^{-3}.

Solution

Poisson's ratio, ν, can be calculated from the following equation:

$$\nu = \frac{1 - \frac{\Delta V}{V_0 \epsilon}}{2}$$

Where ΔV is the change in volume due to the load, V_o is the original volume prior to the load, and ε is the axial strain.

The original volume of the cylinder is:

$$V_o = A * L = \pi \left(\frac{2\ in}{2}\right)^2 \times 8\ ft \left(\frac{12\ in}{1\ ft}\right) = 301.59\ in^3$$

The volume after the load is:

$$V_f = \pi \left(\frac{2\ in - 0.002}{2}\right)^2 \times (8.03\ ft)\left(\frac{12\ in}{1\ ft}\right) = 302.12\ in^3$$

$$\Delta V = 302.12\ in^3 - 301.59\ in^3 = 0.53\ in^3$$

$$v = \frac{1 - \dfrac{0.53\ in^3}{301.59\ in^3 (3.75 \times 10^{-3})}}{2} = 0.27$$

Poisson's ratio for steel is typically between 0.25 and 0.30.

Problem

A 20-foot long section of ductile iron pipe with an outer diameter of 11.26 inches and a wall thickness of 0.52 inches is compressed by a 350-kip axial force. The ductile iron pipe has a modulus of elasticity of 23,000 ksi. Determine the longitudinal stress, the axial strain, and the change in the pipe length due to the force. Assume the material behaves linearly elastically.

Solution

The longitudinal stress, σ, for a system in compression is equal to -P/A, where P is the axial force on the pipe, and A is the cross-sectional area of the pipe. A pipe with an outer diameter of 11.26 inches and a wall thickness of 0.52 inches will have an inner diameter of 10.22 inches (11.26 inches – 2(0.52 inches)). The longitudinal stress is thus:

$$\sigma = \frac{-350\ kips}{\pi \left[\left(\dfrac{11.26\ in}{2}\right)^2 - \left(\dfrac{10.22\ in}{2}\right)^2\right]} = -19.95\ ksi$$

Since the material behaves linearly elastically, the axial strain, ε, may be calculated from Hooke's law, where E is the modulus of elasticity:

$$\epsilon = \frac{\sigma}{E} = \frac{-19.95\ ksi}{23,000\ ksi} = -8.67 \times 10^{-4}$$

The change in pipe length, δ, is calculated from the equation below, where L is the original length of the cylinder:

$$\delta = \epsilon L = -8.67 \times 10^{-4}(20\ ft)\left(\frac{12\ in}{1\ ft}\right) = -0.21\ inches$$

This indicates the pipe length decreases by 0.21 inches.

Problem

A 20-foot long section of ductile iron pipe with an outer diameter of 11.26 inches and a wall thickness of 0.52 inches is compressed by a 350-kip axial force. The axial force results in an axial strain of -8.67x10⁻⁴ and a decrease in pipe length by 0.21 inches. Determine the change in the outer diameter, inner diameter, and wall thickness of the pipe, assuming Poisson's ratio for the material is 0.29.

Solution

First, the lateral strain, ε', is calculated from Poisson's ratio equation, where ν is Poisson's ratio, and ε is the axial strain:

$$\epsilon' = -\nu\epsilon = -0.29(-8.67 \times 10^{-4}) = 2.51 \times 10^{-4}$$

The positive sign of the lateral strain indicates that the compressive force results in the pipe diameters being increased. A pipe with an outer diameter of 11.26 inches and a wall thickness of 0.52 inches will have an inner diameter of 10.22 inches (11.26 inches − 2(0.52 inches)). The increase in the inner diameter is calculated from:
$\Delta d_{inner} = \epsilon' \times d_{inner} = 2.51 \times 10^{-4} \times 10.22\ in = 0.00257\ in$
The increase in the outer diameter is calculated from:
$\Delta d_{outer} = \epsilon' \times d_{outer} = 2.51 \times 10^{-4} \times 11.26\ in = 0.00283\ in$
The increase in the pipe wall thickness is calculated from:
$\Delta t_{wall} = \epsilon' \times t_{wall} = 2.51 \times 10^{-4} \times 0.52\ in = 0.00013\ in$
These values can be verified from:
$\Delta t_{wall} = \dfrac{\Delta d_{outer} - \Delta d_{inner}}{2} = \dfrac{0.00283\ in - 0.00257\ in}{2}$
$\Delta t_{wall} = 0.00013\ in$ ✓

Problem

A 20-foot long section of ductile iron pipe with an outer diameter of 11.26 inches and an inner diameter of 10.22 inches is compressed by a 350-kip axial force. The axial force results in an axial strain of -8.67x10⁻⁴. Calculate the change in volume and the dilatation of the material, assuming Poisson's ratio for the material is 0.29.

Solution
The change in volume, ΔV, can be calculated from the following equation:
$\Delta V = V_o \epsilon (1 - 2v) = AL\epsilon(1 - 2v)$
Where V_o is the original volume prior to the load, A is the cross sectional area of the pipe, L is the length of the pipe, ε is the axial strain, and v is Poisson's ratio. The original volume for this pipe is:
$V_o = AL = \pi \left[\left(\dfrac{11.26\ in}{2}\right)^2 - \left(\dfrac{10.22\ in}{2}\right)^2\right](20ft)\left(12\dfrac{in}{1ft}\right)$
$V_o = 4210.84\ in^3$
The change in material volume is then:
$\Delta V = 4210.84\ in^3\ (-8.67 \times 10^{-4})(1 - 2(0.29))$
$\Delta V = -1.53\ in^3$
The negative sign indicates that the material will decrease in volume as a result of the compressive force.
The dilatation, e, of the material is found from:
$e = \epsilon(1 - 2v)$
$e = (-8.67 \times 10^{-4})(1 - 2(0.29)) = -3.64 \times 10^{-4}$

Problem

A compressive axial force of 1500 lbf acts on a prismatic bar having a cross-sectional area equal to 0.785 in². Imagine a plane cuts through the bar at an angle of 15° (see below). Calculate the stresses on this inclined section.

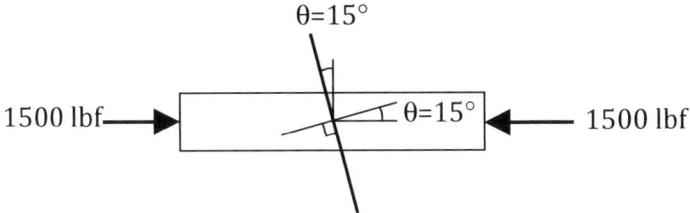

Solution

First, calculate the normal stress σ_x acting on the section from the following equation, where P is the force acting on the bar (negative for compressive forces), and A is the bar's cross-sectional area:

$$\sigma_x = \frac{P}{A} = \frac{-1500 \; lbf}{0.785 \; in^2} = -1910.83 \; lbf/in^2$$

Next, calculate the normal stress σ_θ and the shear stress, τ_θ, taking into account the angle of incline, $\theta=15°$:

$\sigma_\theta = \sigma_x \cos^2\theta = (-1910.83 \; lbf/in^2)(\cos 15)^2$
$\sigma_\theta = -1782.83 \; lbf/in^2$
$\tau_\theta = -\sigma_x(\sin 15)(\cos 15) = -(-1910.83 \; lbf/in^2)(0.25)$
$\tau_\theta = 477.71 \; lbf/in^2$

The normal stress σ_θ is negative (compressive force), and the shear stress is positive (counterclockwise direction).

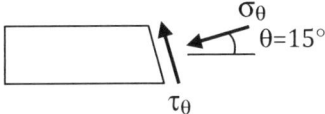

Problem

A compressive axial force acts on a prismatic bar. A plane cutting through the bar at an angle of 15° results in a normal stress σ_θ of -1783 lbf/in² and a shear stress τ_θ of 478 lbf/in² acting on the inclined surface (see below for orientation). The normal stress acting along the x-axis, σ_x, is -1911 lbf/in². Determine and sketch the complete state of stress for $\theta=15°$.

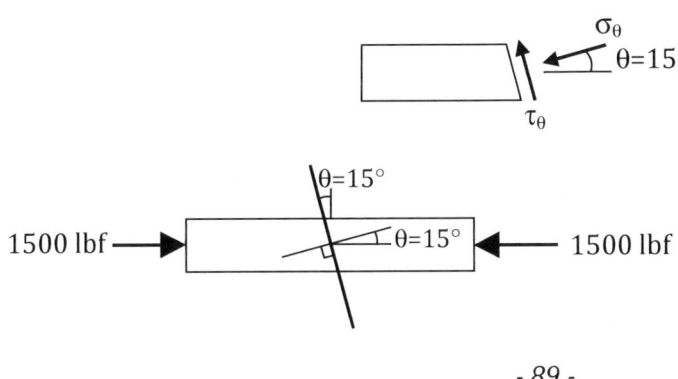

Solution
Face a has the same orientation as that shown in the given information and thus has the same normal and shear stresses as those given. For face c, θ becomes 15°+180° = 195°. For face b, θ becomes 15°-90° = -75°. For face d, θ becomes 15°+90° = 105°. Substituting these θ values into the equations below yield the stresses at each prism face (given σ_x = -1911 lbf/in²).

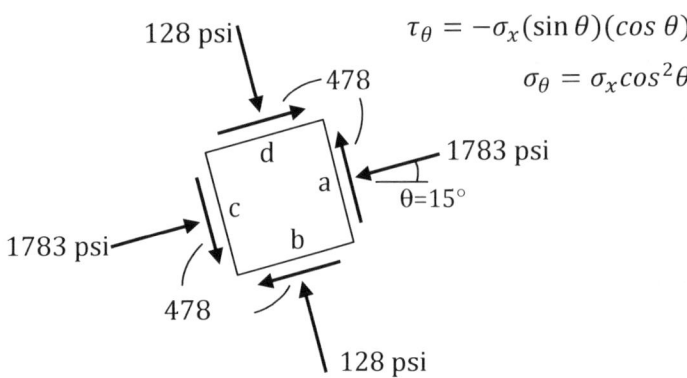

$$\tau_\theta = -\sigma_x(\sin\theta)(\cos\theta)$$
$$\sigma_\theta = \sigma_x\cos^2\theta$$

Problem

Using AISC design methods, determine the allowable strength of a simply supported structural steel W-shape beam in flexure having adequate lateral support and the following properties:
Elastic section modulus, S_x = 128 in³
Plastic section modulus, Z_x = 145 in³
Safety factor for flexure, Ω_b = 1.67
Yield stress, F_y = 50 ksi

Solution
To solve this problem, AISC tables may be used if the beam dimensions were known. Otherwise to calculate the allowable strength, either of the following methods may be used:
Method 1: Calculate the allowable flexural stress, F_b, from the following equation taken from AISC's *Basic Design Values Cards*:
$F_b = 0.66F_y = (0.66)(50\ ksi) = 33\ ksi$
Next, calculate the allowable strength, M_a, using the following equation:
$M_a = F_b S_x = (33\ ksi)(128\ in^3) = 4224\ kip \times inches$
Method No. 2: Calculate the nominal flexural strength, M_n, per AISC's Section F2 design method:
$M_n = F_y Z_x = (50\ ksi)(145\ in^3) = 7250\ kip \times inches$
Next, calculate the allowable strength, M_a, using the following equation:
$$M_a = \frac{M_n}{\Omega_b} = \frac{7250\ kip \times inches}{1.67} = 4341\ kip \times inches$$

Problem

A circular beam made of elastoplastic material has a diameter of 8 inches. Calculate the yield moment and plastic moment in terms of yield stress, σ_y.

Solution
The yield moment, M_y, is calculated from the following equation, where σ_y is the yield stress, I_c is the beam's cross section centroidal moment of inertia, c is the distance from the neutral axis to the extreme surface. I_c for a circular beam is $\pi d^4/64 = \pi(8\ \text{in})^4/64 = 201\ \text{in}^4$. For a circular beam, c is equal to half of the diameter, or 4 inches. The yield moment in terms of yield stress is:

$$M_y = \frac{\sigma_y I_c}{c} = \frac{\sigma_y \times 201\ in^4}{4\ in} = 50.25\ in^3 \times \sigma_y$$

The plastic moment, M_p, is calculated from the following equation, where Z is the plastic modulus, A is the cross-sectional area, $\overline{y_1}$ and $\overline{y_2}$ are the distances to the centroids of the two circle halves (half$_1$ and half$_2$), the latter two being each equal to $2d/3\pi$. The plastic moment in terms of yield stress is:

$$M_p = \sigma_y Z = \frac{\sigma_y A(\overline{y_1} + \overline{y_2})}{2} = 85.3 in^3 \times \sigma_y$$

Modulus of rupture

The modulus of rupture, also known as the flexural strength, is the tensile strength of concrete in flexure. The modulus of rupture is the maximum stress at the beam's surface immediately at failure. It is useful for analyzing potential deflection and cracking in concrete beams. ASTM standards describe tests to determine the flexural strength known as the three point loading test and the center point loading test. The equation for calculating the modulus of rupture or flexural strength, f_r, is shown below:

$$f_r = \frac{Mc}{I}$$

Where M is the maximum bending moment, c is the distance from the neutral axis to the extreme tension fiber, and I is the moment of inertia. For normal-weight concrete, the following equation is typically used to calculate the modulus of rupture:

$f_r = 7.5\sqrt{f_c'}\ (U.S.Customary\ System\ units)$

For light-weight concrete, the modulus of rupture used in design is typically equal to 75% of the calculated value.

Problem

A concrete mix design has the proportions 1:2.3:2.9 by weight. Each sack of cement requires 6.1 gallons of water and weighs 94 lbf. Assume the solid densities of the cement, fine aggregate and course aggregate are 196 lbf/ft³, 163 lbf/ft³, and 170 lbf/ft³, respectively. Calculate the concrete yield per sack of cement.

Solution
The concrete mix proportions 1:2.3:2.9 by weight indicates the ratio of cement to fine aggregate (i.e., sand) to coarse aggregate (i.e., stone) by weight per one unit volume of concrete.

Component	Ratio	Weight per Cement Sack (lbf)	Solid Density (lbf/ft³)	Total Solid Volume (ft3/sack)
Cement	1	1(94) = 94	196	94/196 = 0.480
Fine Aggregate	2.3	2.3(94) = 216.2	163	216.2/163 = 1.326
Coarse Aggregate	2.9	2.9(94) = 272.6	170	272.6/170 = 1.604
Water			7.48 gal/ft³	6.1/7.48 = 0.816
			TOTAL	4.226

The solid yield per cement sack is 4.226 ft³ of concrete.

Problem

A mix design has the following properties:
Mix design proportions: 1:2.3:2.9 by weight
Water required per cement sack: 6.1 gallons
Weight of each cement sack: 94 lbf
Solid yield of concrete per sack: 4.23 ft³

Determine how many sacks are required to make 65 cubic yards of concrete. Also, calculate the amounts of fine aggregate, coarse aggregate and water needed to make 65 cubic yards of concrete.

Solution
The number of sacks of cement required to yield 65 cubic yards of concrete is:
$$No.\,of\,sacks = \frac{65\,yd^3\left(\frac{27\,ft^3}{1\,yd^3}\right)}{4.23\,ft^3/sack} = 414.9\,sacks = 415\,sacks$$
The concrete mix proportions 1:2.3:2.9 by weight indicates the ratio of cement to fine aggregate to coarse aggregate by weight per one unit volume of concrete.
The amount of fine aggregate needed is:
$Weight_{fines} = 2.3(Weight_{cement})$
$Weight_{fines} = 2.3(415\,sacks)\left(94\frac{lbf}{sack}\right) = 89{,}723\,lbf$
The amount of coarse aggregate needed is:
$Weight_{coarse} = 2.9(Weight_{cement})$
$Weight_{coarse} = 2.9(415\,sacks)\left(94\frac{lbf}{sack}\right) = 113{,}129\,lbf$
The amount of water needed is:
$$Volume_{water} = (415\,sacks)\left(\frac{6.1\,gal}{sack}\right) = 2531.5\,gal$$

Cathodic protection

Cathodic protection is a means of protecting the steel within steel-reinforced concrete from corroding. Chloride (in the form of calcium chloride) is frequently added to concrete to increase the concrete's early strength. The chloride, however, can cause the reinforcing steel within the concrete to corrode, thus decreasing the strength of the structure. Although the ACI code limits the chloride concentrations for several applications, additional measures, such as cathodic protection, are sometimes a cost-effective necessity. Cathodic protection involves a continuous, low-power consuming electrical current supplied from an external power source that is distributed through an anode. The electrical current makes its way to the reinforcing steel within concrete via the concrete pores. The steel acts as a cathode, while the water and salt within the concrete pores act as the electrolyte. Cathodic protection substantially lowers the corrosion potential of the steel while being relatively simple to install and maintain.

Problem

Using both the ASD and LRFD methods, select an ASTM A992 (F_y=50 ksi) steel W 12 shape to support an axial dead load of 175 kips and a live load of 350 kips. The column is 18 feet long and is pinned at both ends.

Solution
First calculate the total load to be carried:
$P(ASD) = Dead\ Load + Live\ Load$
$P(ASD) = 175\ kips + 350\ kips = 525\ kips$
$P(LRFD) = 1.2 \times Dead\ Load + 1.6 \times Live\ Load$
$P(LRFD) = 1.2(175\ kips) + 1.6(350\ kips) = 770\ kips$

For columns pinned at both ends, the recommended design value of the effective length factor, K, is 1.0. Determine the value of KL, where L is the unbraced column length. KL is equal to (1.0)(18 ft)=18 ft. Using AISC's W 12 shape steel column design tables (F_y=50 ksi), start with KL=18 ft, and look for the ASD P_n/Ω_c value that is closest to but exceeds 525 kips and for the LRFD $\varphi_c P_n$ value that is closest to but exceeds 770 kips. A W 12x87 shape is the lightest weighing W 12 shape that can handle 525 kips (ASD method) and 770 kips (LRFD method).

*Table values based on the 13th Edition of AISC's *Steel Construction Manual*.

Problem

An ASTM A992 (F_y=50 ksi) W 10x88 steel column is 18 feet long and is braced in the middle in the y-direction and pinned at both ends. Determine the column's maximum allowable axial load using the ASD method.

Solution
First calculate the ratio r_x/r_y using the column properties found in AISC's *Steel Construction Manual* (13th Edition). For a W10x88 steel column, r_x=4.53 inches and r_y=2.63 inches. The ratio r_x/r_y is then equal to 1.72. Next, calculate the slenderness ratio about the y-axis:

$$SR_y = \frac{KL_y}{r_y} = \frac{(1.0)(9\ ft)(12in/1ft)}{2.63\ inches} = 41.06$$

Then calculate the slenderness ratio about the x-axis:

$$SR_x = \frac{KL_x}{r_x} = \frac{(1.0)(18\ ft)(12in/1ft)}{4.53\ inches} = 47.68$$

Since the slenderness ratio about the x-axis is larger than that about the y-axis, the former governs.

Calculate the equivalent effective length about the major axis with respect to the y-axis:

$$(KL_y)_{equiv} = \frac{KL_x}{(r_x/r_y)} = \frac{(1.0)(18\ ft)}{1.72} = 10.47\ ft$$

From the AISC W shape design tables for a W 10x88 column with an effective length of 10.47 ft, the allowable axial load is 656.13 kips.

Problem

Using the ASD method, select an ASTM A992 (F_y=50 ksi) W shape steel column to support a total axial load of 540 kips. The column is 20 feet long (unbraced) and is fixed at both ends.

Solution

For columns that are fixed at both ends, the recommended design value of the effective length factor, K, is 0.65. The value of the effective length, KL, is then equal to (0.65)(20 feet) = 13 feet. Using AISC's W 12 shape steel column design tables (F_y=50 ksi), start with KL=13 feet, and look for the ASD P_n/Ω_c value (ASD method) that is closest to but exceeds 540 kips. From the tables, the following W shape columns could work:

a) W 10x88 shape column with a P_n/Ω_c value of 600 kips with a weight of 88 pounds per linear foot.

b) W 12x79 shape column with a P_n/Ω_c value of 574 kips with a weight of 79 pounds per linear foot.

c) W 14x90 shape column with a P_n/Ω_c value of 697 kips with a weight of 90 pounds per linear foot.

It is likely that the best selection among these is the lightest column, which would be the W 12x79 shape.

*Table values based on the 13th Edition of AISC's *Steel Construction Manual*.

Reinforcing steel bar designation

Reinforcing steel bar (also known as rebar) is commonly used to increase the tensile strength of concrete and masonry structures. Deformed rebar is rebar that exhibits a ribbed surface pattern to increase the bonding between the rebar and wet concrete. In the U.S., both imperial and metric system rebar designation are used. Using the imperial designation system, rebar with the number "4" stamped on it indicates that the nominal diameter of the bar (excluding any deformation) is 4/8" or 0.5." Similarly, rebar stamped with the number "6" (imperial system) indicates the nominal diameter is 6/8" or 0.75". This trend (one-eighth inch diameter increments) goes up through #8 bar. The #4 bar using the imperial system is equivalent to the #13 bar using the metric system. The metric system designation is a "soft" nominal diameter measurement in millimeters. The #13 bar (metric system) has a nominal diameter of 12.7 mm. The #19 bar using the metric system (equivalent to the imperial system #6 bar) has a nominal diameter of 19.05 mm.

Problem

A L8x6x1/2 angle support is subjected to a bending moment at the vector location M shown below. The angle measurement between axis 2-2 and axis 3-3 is 29.2°. Given: axis 3-3 is the y-axis and axis 4-4 is the z-axis. Determine the principal moments of inertia Iy and Iz. Also determine the angle between the z-axis and the neutral axis.

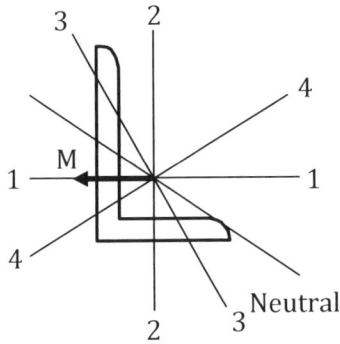

Solution
From tabulated information for L8×6×1/2 angle supports, the area, A, is 6.75 in² and the minimum radius of gyration, r_{min}, is 1.30 in. The principal moment of inertia, I_y ($I_{3\text{-}3}$) is then:
$I_y = A \times r_{min}^2 = (6.75 \ in^2)(1.30 \ in)^2 = 11.4 \ in^4$
The principal moment of inertia, I_z ($I_{4\text{-}4}$), can be determined from the following equation, where $I_{1\text{-}1}$ and $I_{2\text{-}2}$ are tabulated values of 44.3 in⁴ and 21.7 in⁴, respectively:
$I_z = I_{1-1} + I_{2-2} - I_y = 44.3 \ in^4 + 21.7 \ in^4 - 11.4 \ in^4$
$I_z = 54.6 \ in^4$
The angle between axis 4-4 (the z-axis) and the neutral axis, α, is calculated from:
$$\tan \alpha = \frac{I_z}{I_y} \tan \theta$$
Where θ is the angle between axis 2-2 and axis 3-3 (given). Thus:
$$\tan \alpha = \frac{54.6 \ in^4}{11.4 \ in^4} \tan 29.2° = 2.68$$
$$\alpha = 69.5°$$

Problem

A simply supported 25-ft long reinforced concrete beam (cross-sectional area equal to 216 in²) has a dead load of 1.8 kips per foot and a live load of 0.5 kips per foot (beam's dead weight not included). Determine the maximum moment, assuming no wind loading.

Solution
The total load will first need to be determined. Steel-reinforced concrete has a typical weight of 150 lbf/ft³ and should be factored into the dead load. The dead load is:

$$Dead \ Load = 1.8 \frac{kips}{ft} + \frac{\left(150 \frac{lbf}{ft^3}\right)(216 \ in^2)}{\left(\frac{12 \ in}{1 \ ft}\right)^2 \left(1000 \frac{lbf}{kip}\right)} = 2.0 \ kips/ft$$

For reinforced concrete, the design load combination used is:
Total Load = 1.2 × Dead Load + 1.6 × Live Load
Total Load = 1.2(2.0 kips/ft) + 1.6(0.5 kips/ft)
Total Load = 3.2 kips/ft

The maximum moment is located at the center of a uniformly loaded beam. The equation to calculate the moment at the center of the beam is:

$$M_{max} = \frac{\text{Total Load } (L^2)}{8} = \frac{3.2 \text{ kips/ft } (25 \text{ ft})^2}{8}$$

$$M_{max} = 250 \text{ kip} \times \text{ft}$$

Problem

A simply supported uniformly loaded 25-ft long reinforced concrete beam's cross section is shown below. Determine the maximum allowable tension reinforcing steel area (without the use of compression steel) if $f_c' = 3,950$ lbf/in² and $f_y = 54,000$ lbf/in².

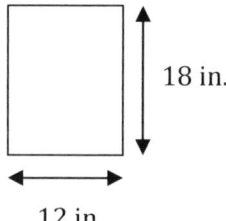

18 in.

12 in.

Solution
First calculate the balanced steel area, A_{sb}, which is the area value of tensile steel at which steel yielding will occur at the same instant that the concrete will compressively crush. For rectangular beams, the balanced steel area can be calculated from the following equation:

$$A_{sb} = bd\left(\frac{0.85\beta_1 f_c'}{f_y}\right)\left(\frac{87,000}{f_y + 87,000}\right)$$

Where b is the beam width, d is the beam depth and β_1 is a factor equal to 0.85 if $f_c' \leq 4,000$ lbf/in². The equation becomes:

$$A_{sb} = (12 \text{ in})(18 \text{ in})\left(\frac{0.85(0.85)3950 \frac{lbf}{in^2}}{54,000 \frac{lbf}{in^2}}\right)\ldots$$

$$\ldots\left(\frac{87,000}{54,000 \text{ lbf/in}^2 + 87,000}\right) = 7.04 \text{ in}^2$$

The maximum allowable tensile reinforcing steel, $A_{s,max}$, is then calculated from the following:

$$A_{s,max} = 0.75 A_{sb} = 0.75(7.04 \text{ in}^2) = 5.28 \text{ in}^2$$

Problem

*A simply supported uniformly loaded 25-ft long reinforced concrete beam has a maximum moment, M_{max}, equal to 285 kip*ft at the beam's center. The beam's cross section is shown below. Calculate the required area of concrete to balance the steel force at yield if $f_c' = 3,950$*

lbf/in² and f_y = 54,000 lbf/in². Assume the distance from the centroid of the compression zone to the most compressed fiber λ = 0.1d.

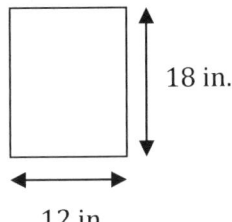

12 in.

Solution
First, determine the amount or area of reinforcing steel required, $A_{s,req}$, for this beam by using the following equation:

$$A_{s,req} = \frac{M_{max}}{\phi f_y (d - \lambda)}$$

Where ϕ is the strength reduction factor, d is the beam depth, and λ is the distance from the centroid of the compression zone to the most compressed fiber. The strength reduction factor for beams in flexure is 0.9. λ is equal to 0.1d or 1.8 inches. The equation becomes:

$$A_{s,req} = \frac{(285\ kip \times ft)(12\ in/1\ ft)}{(0.9)\left(54{,}000\ \frac{lbf}{in^2}\right)\left(\frac{1\ kip}{1000\ lbf}\right)(18\ in - 1.8\ in)}$$

$A_{s,req} = 4.34\ in^2$

Then calculate the required area of concrete to balance the steel force at yield:

$$A_c = \frac{f_y \times A_{s,req}}{0.85 f_c'} = \frac{54{,}000\ \frac{lbf}{in^2}(4.34\ in^2)}{0.85(3950\ lbf/in^2)} = 69.8\ in^2$$

Problem

A simply supported uniformly loaded 25-ft long reinforced concrete beam's cross section is shown below. Calculate the location of the compression zone centroid and the design moment capacity, given the following:
Area of reinforcing steel required, $A_{s,req}$ = 4.34 in²;
Req. concrete area to balance steel force at yield, A_c = 69.8 in²
f_c' = 3,950 lbf/in² and f_y = 54,000 lbf/in²
Initial assumed value of λ = 0.1d
Strength reduction factor, ϕ, = 0.9

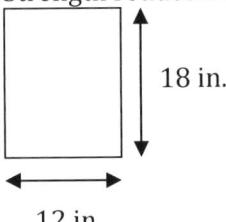

12 in.

Solution
The first step is to determine the thickness of the compression zone. Since the beam is 12 inches (b) wide, the compression zone will be 12 inches wide. For a rectangular beam, the thickness of the compression zone, t, can then be determined from:

$$t = \frac{A_c}{b} = \frac{69.8 \ in^2}{12 \ in} = 5.82 \ in$$

The centroid of the compression zone for a rectangular-shaped beam is then calculated from:

$$\lambda = \frac{t \times b \times \frac{t}{2}}{A_c} = \frac{5.82 \ in (12 \ in) \left(\frac{5.82 \ in}{2}\right)}{69.8 \ in^2} = 2.91 \ in$$

The nominal moment capacity can be calculated from:
$$M_n = A_{s,req} f_y (d - \lambda)$$
$$M_n = 4.34 \ in^2 \left(\frac{54,000 \ lbf}{in^2}\right) \left(\frac{kip}{1000 \ lbf}\right) (18 \ in - 2.91 \ in)$$
$$M_n = 3536.5 \ kip \times in \left(\frac{1 ft}{12 in}\right) = 294.71 \ kip \times ft$$

The design moment capacity is equal to ϕM_n = (0.9)(294.71 kip*ft) = 265.24 kip*ft.

Problem

One-foot wide beams, spaced 10 feet apart on centers support a reinforced concrete slab. Calculate the required thickness, the slab weight, and the total factored load of this one-way slab (one end continuous) having the following properties:
Dead load of 18 lbf/ft² , not including slab's weight
Live load of 43 lbf/ft²

Solution
The span between beams is given as 10 feet on beam center. The clear span between the beams is 10 feet – 2(6 in)(1 ft/12 in)= 9 feet. The slab thickness required by ACI 318-05 for a one end continuous slab is 1/24 of the clear span length. The required slab thickness is then:

$$h \geq \frac{1}{24} \times length_{clear \ span} = \frac{1}{24}(9 \ ft)\left(\frac{12 in}{1 ft}\right) = 4.5 \ in$$

The required slab thickness (rounding up) is thus 5 inches.
Assuming the weight of reinforced concrete is 150 lbf/ft, the reinforced concrete slab weight is:

$$Weight_{slab} = \left(150 \frac{lbf}{ft^3}\right)(5 \ in)\left(\frac{1 ft}{12 in}\right) = 62.5 \ lbf/ft^2$$

The total factored load is then calculated from:
$$Total \ Load = 1.2 \times Dead \ Load + 1.6 \times Live \ Load$$
$$Total \ Load = 1.2\left(62.5 \frac{lbf}{ft^2} + 18 \frac{lbf}{ft^2}\right) + 1.6\left(43 \frac{lbf}{ft^2}\right)$$
$$Total \ Load = 165 \ lbf/ft^2$$

Problem

One-foot wide beams, spaced 10 feet apart on centers (clear span of 9 feet) support a reinforced concrete one-way slab (one end continuous). In addition, the slab is simply supported at the far ends and supported with integral supports every 30 feet. Calculate the positive factored moments for this slab having the following:
Total factored load, w_u, of 165 lbf/ft² (0.165 kip/ ft²)

Solution
ACI 318-05 provides allowable moment coefficients, C_1, via the following equation: M = $C_1 w_u L^2$, where w_u is the factored load and L is the clear span length. For simply supported end spans, $C_1 = 1/11$. For interior spans, $C_1 = 1/16$. For end spans with built-in support (integral support), $C_1 = 1/14$.

Calculate the moments assuming a 1 foot wide strip. The positive moment of the end span with the simple support is:

$$M_{e,ss} = \frac{(w_u) span_{clear}^2}{11} = \frac{\left(0.165 \frac{kip}{ft}\right)(9\ ft)^2}{11} = 1.22\ kip \times ft$$

The positive moment of the interior span is:

$$M_i = \frac{(w_u) span_{clear}^2}{16} = \frac{\left(0.165 \frac{kip}{ft}\right)(9\ ft)^2}{16} = 0.84\ kip \times ft$$

The positive moment of the end span with the integral support is:

$$M_{e,is} = \frac{(w_u) span_{clear}^2}{14} = \frac{\left(0.165 \frac{kip}{ft}\right)(9\ ft)^2}{14} = 0.95\ kip \times ft$$

Problem

One-foot wide beams, spaced 10 feet apart on centers (clear span of 9 feet) support a reinforced concrete one-way slab (one end continuous). In addition, the slab is simply supported at the far ends and supported with integral supports every 30 feet. Calculate the negative factored moments for this slab having the following:
Total factored load, w_u, of 165 lbf/ft² (0.165 kip/ ft²)

Solution
ACI 318-05 provides allowable moment coefficients, C_1, via the following equation: M = $C_1 w_u L^2$, where w_u is the factored load and L is the clear span length. For exterior built-in supports (integral supports) faces that are cross beams, $C_1 = 1/24$. For exterior built-in supports at faces that are columns, $C_1 = 1/16$. For interior support faces, $C_1 = 1/11$ (all cases).

Calculate the moments assuming a 1 foot wide strip. The exterior negative moment at face of cross beam is:

$$M_{e,cbf} = \frac{(w_u) span_{clear}^2}{24} = \frac{\left(0.165 \frac{kip}{ft}\right)(9\ ft)^2}{24} = 0.56\ kip \times ft$$

The exterior negative moment at face of column is:

$$M_{e,cf} = \frac{(w_u) span_{clear}^2}{16} = \frac{\left(0.165 \frac{kip}{ft}\right)(9\ ft)^2}{16} = 0.84\ kip \times ft$$

The interior negative moment at support face:

$$M_{isf} = \frac{(w_u) span_{clear}^2}{11} = \frac{\left(0.165 \frac{kip}{ft}\right)(9\ ft)^2}{11} = 1.22\ kip \times ft$$

Problem

One-foot wide beams, spaced 10 feet apart on centers (clear span of 9 feet) support a reinforced concrete one-way slab (one end continuous). In addition, the slab is simply

supported at the far ends and supported with integral supports every 30 feet. Calculate the maximum factored shear (shear coefficient C_2=1.15) and the shear capacity of this slab having the following properties:
Total factored load, w_u, of 165 lbf/ft² (0.165 kip/ft²); Effective depth (shear calculation) = 3.5 inches; f_c' = 3950 lbf/in²

Solution
The maximum factored shear is calculated from:
$$V_{F,max} = C_2 \frac{w_u L}{2}$$
Where C_2 is the ACI 318-05 shear coefficient, w_u is the factored load and L is the clear span length. Assuming a 1-ft wide strip:
$$V_{F,max} = 1.15 \frac{\left(0.165 \frac{kip}{ft^2}\right)(1\ ft)(9\ ft)}{2} = 0.85\ kip$$
The shear capacity is calculated from:
$$V_c = 2db_w\sqrt{f_c'}$$
Where d is the effective depth, b_w is the beam width and f_c' is the proposed specified design strength in pounds per square inch.
$$V_c = 2(3.5\ in)(12\ in)(1\ kip/1000\ lbf)\sqrt{\left(3950 \frac{lbf}{in^2}\right)}$$
$V_c = 5.28\ kips$
Factored shear capacity = ϕV_c = 0.75(5.28 kips) = 3.96 kips
3.96 kips > 0.85 kip ✓

Problem

An 18-inch square column is supported by a 132-inch square reinforced concrete footing. For an effective footing depth of 12.5 inches, determine the punching shear stress given the following:
Total factored load, w_u, of 330 kips

Solution
To calculate the punching shear stress, the area that resists the punching shear, A_p, and the upward force from the soil pressure that reduces the punching shear force, R, will first need to be calculated:
$$A_p = 2d(b_1 + b_2)$$
Where d is the effective footing depth, and for a square column, b_1 and b_2 are both equal to the sum of the column width and the footing depth. The area resisting punching shear is then:
$$A_p = 2(12.5\ in)(2)(18\ in + 12.5\ in) = 1525\ in^2 = 10.6\ ft^2$$
R is calculated from:
$$R = w_u \left(\frac{b_1 b_2}{A_f}\right)$$
Where A_f is the area of the footing. R is then:
$$R = 330\ kips \left(\frac{(30.5\ in)^2}{(132\ in)^2}\right) = 17.6\ kips$$
The punching shear stress is then calculated from:

$$v_u = \frac{w_u - R}{A_p} = \frac{330 \text{ kips} - 17.6 \text{ kips}}{10.6 \text{ } ft^2} = 29.5 \text{ } kip/ft^2$$

Transportation

Horizontal curve abbreviations

T refers to the semi-tangent distance on a horizontal curve. The endpoints of the semi-tangent are the point of curvature, PC, and the tangent intersection point vertex, PI, or the point of tangency, PT, and the tangent intersection point vertex, PI. PC is the point of curvature. On a horizontal curve, the PC is the point where the curve begins and the back tangent ends. PT is the point of tangency and is the point on a horizontal curve where the curve ends and the forward tangent begins. R is the radius of the curve. POC is the abbreviation given to any point on the curve. POT is the abbreviation given to any point on the tangent (either the back or forward tangent). PI is the point that the back and forward tangents intersect. $POST$ is the abbreviation for any point on the semi-tangent. C refers to the long chord distance, also referred to as LC. The long chord is the distance between the point of curvature, PC, and the point of tangency, PT. M is the distance from the midpoint of the curve, MPC, to the midpoint of the long chord.

Problem

Label the following components on the horizontal curve: back tangent, forward tangent, semi-tangent, point of intersection of back and forward tangents, length of curve, curve midpoint, middle ordinate, external distance, long chord, curve radius, interior angle, point of curvature, and point of tangency.

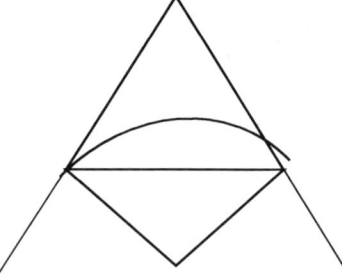

Solution

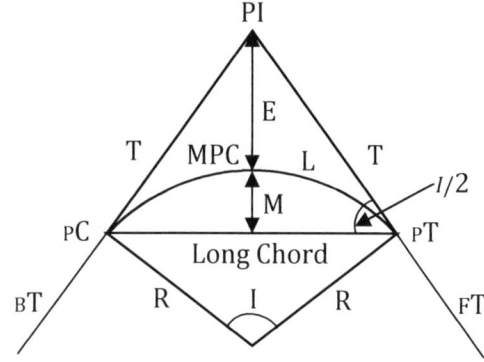

BT (back tangent), FT (forward tangent), T (semi-tangent), PI (point of intersection of back and forward tangents), L (length of curve), MPC (midpoint of curve), M (middle ordinate), external distance (E), the long chord (C), R (curve radius), I (interior angle), PC (point of curvature), and PT (point of tangency).

Problem

A horizontal curve has an interior angle of 10.2°. Determine the locations of the PC and PT stations if the PI station is 32+47.43 and the curve radius is 1,785.51 feet.

Solution
The PT station can be determined by adding the length of the curve to the PC station. The PC station can be determined by subtracting the semi-tangent distance from the PI station. The semi-tangent distance, T, is calculated from the following:

$$T = R \times tan\left(\frac{I}{2}\right)$$

Where R is the curve radius and I is the interior angle.

$$T = 1785.51 \: ft \times tan\left(\frac{10.2°}{2}\right) = 159.35 \: ft$$

The length of the curve, L, is calculated from the following:

$$L = RI\left(\frac{2\pi}{360°}\right) = (1785.51 \: ft)(10.2°)\left(\frac{2\pi}{360°}\right) = 317.86 \: ft$$

The PC station is then at:
$$sta \: PC = sta \: PI - T = (32 + 47.43) - 159.35 \: ft$$
$$sta \: PC = 30 + 88.08$$

The PT station is then at:
$$sta \: PT = sta \: PC + L = (30 + 88.08) + 317.86 \: ft$$
$$sta \: PT = 34 + 05.94$$

Problem

A horizontal curve has an interior angle of 10.2°. Calculate the long chord length, the middle ordinate, and the external distance if the curve radius is 1,785.51 feet.

Solution
The long chord length, C, can be calculated from the following:

$$C = 2R sin\left(\frac{I}{2}\right)$$

Where R is the curve radius and I is the interior angle.

$$C = (2)(1785.51 \: ft) \times sin\left(\frac{10.2°}{2}\right) = 317.44 \: ft$$

The middle ordinate distance, M, or the distance between the midpoints of the curve and long chord, can be calculated from:

$$M = R\left(1 - cos\left(\frac{I}{2}\right)\right) = (1785.51 \: ft)\left(1 - cos\left(\frac{10.2°}{2}\right)\right)$$
$$M = 7.07 \: ft$$

Alternatively, the following equation could have been used:

$$M = \left(\frac{C}{2}\right) tan\left(\frac{I}{4}\right)$$

The external distance, E, or the distance between the PI and the curve midpoint is calculated from:

$$E = R\left(\sec\left(\frac{I}{2}\right) - 1\right) = (1785.51\ ft)\left(\sec\left(\frac{10.2°}{2}\right) - 1\right)$$
$$E = 7.10\ ft$$

Problem

Define tangent offset and tangent distance for a circular curve and solve for each if the subtended arc radius is 1406 feet and the internal angle, β, is 37°.

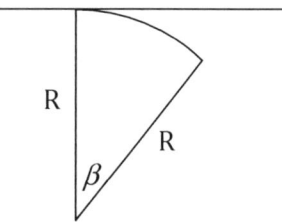

Solution

If the tangent line to the beginning of a circular curve is extended, the tangent offset is the perpendicular distance from that extended tangent to the curve (y below). The tangent distance is the distance from the beginning of the curve to the point the tangent offset meets the extended tangent line (x below).

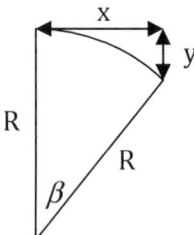

The tangent distance, x, can be calculated from:
$x = R \sin\beta = 1406\ ft (\sin 37°) = 846.15\ ft$
The tangent offset, y, can be calculated from:
$y = R(1 - \cos\beta) = (1406\ ft)(1 - \cos 37°) = 283.12\ ft$

Problem

Solve for the distances FG and GH if the deflection angle is 11° and the curve radius is 1500 ft.

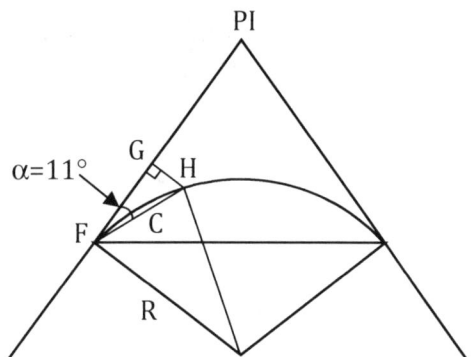

Solution
The deflection angle, α, is the angle between the tangent and the chord. The angle at which the two curve radii meet is equal to two times the deflection angle, as shown below:

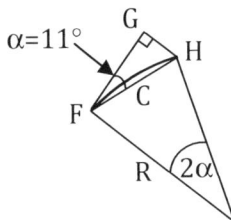

The chord distance, C (or FH), may be calculated using the deflection angle by the following equation:
$C = FH = 2R \sin \alpha = 2(1500\ ft)(\sin 11°) = 572.43\ ft$
The distance FG (the tangent distance) can be calculated from right angle equations:
$FG = FH \cos \alpha = 572.43\ ft(\cos 11°) = 561.91\ ft$
The distance GH (the tangent offset) can be calculated from:
$GH = FH \sin \alpha = 572.43\ ft(\sin 11°) = 109.22\ ft$

Problem

Calculate the chord distance FJ and the chord offset JH for a horizontal curve having a deflection angle of 15° a curve radius of 1250 ft, and an interior angle of 104°.

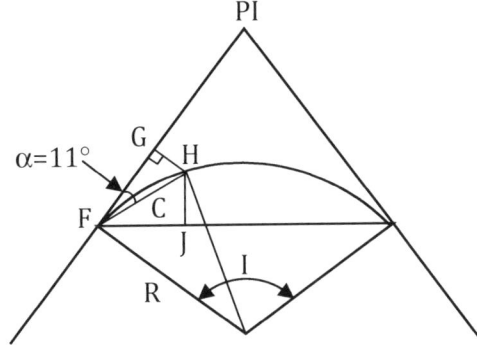

Solution
A horizontal curve may be created by the chord offset method, which locates points on the curve at offsets to distances along the main chord. The angle between the chord FH and the chord distance FJ is equal to $(I/2) - \alpha = (104°/2) - 15° = 37°$ (shown below), where I is the interior angle and α is the deflection angle.

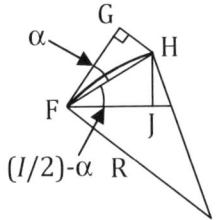

$FH = 2R \sin \alpha = 2(1250\ ft)(\sin 15°) = 647.05\ ft$
The chord distance FJ and the chord offset JH are then:

$$FJ = FH * \cos\left(\frac{I}{2} - \alpha\right) = 647.05 \, ft \times \cos\,(37°) = 516.76 \, ft$$
$$JH = FH * \sin\left(\frac{I}{2} - \alpha\right) = 647.05 \, ft \times \sin\,(37°) = 389.40 \, ft$$

Problem

A horizontal curve having an interior angle of 30.3° is staked every 100 feet. The central angle for each 100 ft station is 5.73° as shown below, except for the last section, at which the curve ties into the forward tangent. Calculate a) the curve radius, b) the chord length for each 100 ft arc, c) the central angle of the final chord and d) the length of the final chord.

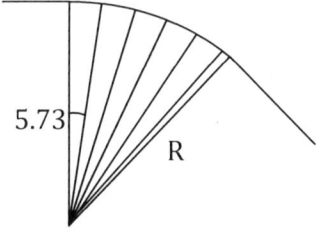

Solution
a) If the central angle, β, and the length of the arc are known, the curve radius can be calculated from:
$$R = \frac{(360°)L_{arc}}{2\pi\beta} = \frac{(360°)(100 \, ft)}{2\pi(5.73°)} = 999.93 \, ft$$
b) The chord length, C, for a central angle of 5.73° and a radius of 1000 ft can be calculated from:
$$C = 2R \sin\left(\frac{\beta}{2}\right) = 2(999.93 \, ft)\sin\left(\frac{5.73°}{2}\right) = 99.96 \, ft$$
c) The central angle of the final chord is calculated by dividing the interior angle, 30.3°, by the central angles of the 100' arcs: (30.3°/5.73°) = 5.28. The central angle of the final chord is then: (30.3° - 5(5.73°)) = 1.65°.
d) The length of the final chord can be calculated from:
$$C_{final} = 2R \sin\left(\frac{\beta}{2}\right) = 2(999.93 \, ft)\sin\left(\frac{1.65°}{2}\right) = 28.79 \, ft$$

Problem

For the horizontal compound curve below, calculate the semi-tangent T_A, the central angle β_1, and
radius R_1 given that radius R_2 = 515 feet, the central angle β_2 = 47°, tangent T_B = 873.45 ft, and the interior angle, I = 72°.

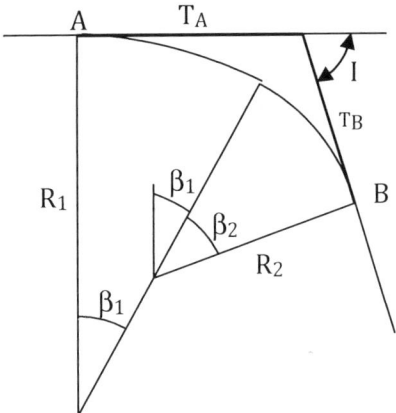

Solution
The deflection angle β_1 is calculated from the following equation:
$\beta_1 = I - \beta_2 = 72° - 47° = 25°$

Radius R_1 is calculated from the following equation:
$$R_1 = R_2 + \left[\frac{(T_B \sin I - R_2(1 - \cos I))}{(1 - \cos \beta_1)}\right]$$
$$R_1 = R_2 + \left[\frac{(873.45\, ft(\sin 72°) - 515\, ft(1 - \cos 72°))}{(1 - \cos 25°)}\right]$$
$$R_1 = 515\, ft + 5068.13\, ft = 5583.13\, ft$$

The tangent T_A can then be calculated from:
$$T_A = \frac{[R_2 - R_1 \cos I + (R_1 - R_2) \cos \beta_2]}{\sin I}$$
$$T_A = \frac{[515 - 5583.13 \cos 72° + (5583.13 - 515) \cos 47°]}{\sin 72°}$$
$$T_A = 2361.77\, ft$$

Vertical curve abbreviations

BVC refers to the beginning of the vertical curve and is the point the back tangent ends and the vertical curve begins.

EVC is the end of the vertical curve, the point the vertical curve ends and the forward tangent begins.

V refers to the vertex, or the point at which the extended back and forward tangents meet.

G_1 is the percent grade at which the stationing begins (from the BVC to the vertex).

G_2 is the percent grade at which the stationing heads (from the vertex to the EVC).

A is the change in gradient, which is calculated by subtracting G_1 from G_2.

R is the rate of change in gradient per station. The change in gradient is divided by the number of stations to obtain R.

L is the length of the vertical curve (horizontal projection, from the beginning of the vertical curve (BVC) to the end of the vertical curve (EVC).

M refers to the middle ordinate; the perpendicular distance between the vertex (intersection) of the back and forward tangents to the curve.

TP refers to a turning point, or a point at which the slope of the curve is equal to zero.

Problem

Label the following components on the vertical crest and sag curves: beginning of vertical curve, end of vertical curve, middle ordinate, vertex, G_1, G_2, turning point, and the length of the curve.

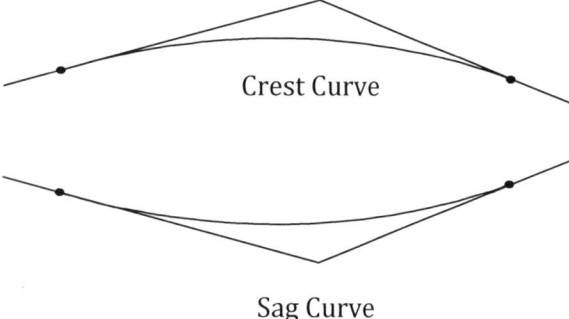

Crest Curve

Sag Curve

<u>Solution</u>
V = vertex, BVC = curve beginning, EVC = curve end, TP = turning point, L = curve length, and M = middle ordinate.

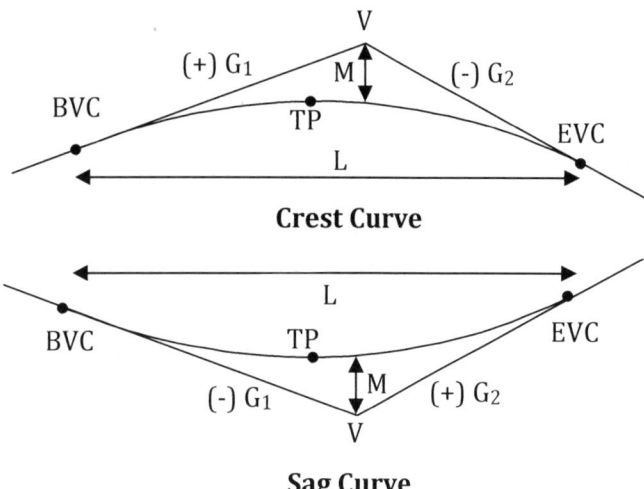

Crest Curve

Sag Curve

Problem

A 600-ft crest vertical curve has a vertex elevation of 1236.57 feet. The BVC and EVC elevations are 1232.98 feet and 1234.28 feet, respectively. Calculate G_1, G_2, and the rate of grade change.

Solution
The horizontal projection of the curve length, L, is equal to 600 ft/(100 ft/sta) = 6 stations. The equations to calculate the gradient of the back tangent (G_1) and forward tangent (G_2) are:

$$Elev_{BVC} = Elev_{Vertex} - G_1\left(\frac{L}{2}\right)$$

$$Elev_{EVC} = Elev_{Vertex} - G_2\left(\frac{L}{2}\right)$$

Solving for G_1 in the first equation:

$$G_1 = \left(\frac{2}{L}\right)(Elev_{Vertex} - Elev_{BVC})$$

$$G_1 = \left(\frac{2}{6\ sta}\right)(1236.57ft - 1232.98ft) = 1.20\%$$

Solving for G_2 in the second equation:

$$G_2 = \left(\frac{2}{L}\right)(Elev_{Vertex} - Elev_{EVC})$$

$$G_2 = \left(\frac{2}{6\ sta}\right)(1236.57ft - 1234.28\ ft) = 0.76\%$$

Because the curve is a crest curve, G_2 = -0.76%.

Problem

A 600-ft crest vertical curve has a vertex elevation of 1236.57 feet. The following information is known: the elevation of the BVC, located at station 55+00, is 1232.98 feet, G_1 = 1.20% (1.20 ft/station) and G_2 = -0.76% (0.76 ft/station). Calculate the rate of grade change and the elevation at station 57+00.

Solution
The horizontal projection of the curve length, L, is equal to 600 ft/(100 ft/sta) = 6 stations. The rate of change in grade per station, R, is calculated from:

$$R = \frac{G_2 - G_1}{L} = \frac{-0.76\% - 1.20\%}{6\ stations} = -0.4933\%\ per\ station$$

This is also equal to -0.4933 ft/station².
The equation to find the distance, in stations, to a point, x, on the vertical curve is determined from the following:

$$Elev_x = \left(\frac{R}{2}\right)x^2 + G_1 x + Elev_{BVC}$$

Station 57+00 is 200 feet or 2 stations away from the BVC at station 55+00.

$$Elev_{57+00} = \left(\frac{-0.4933\frac{ft}{sta^2}}{2}\right)(2\ sta)^2 + \cdots$$

$$\cdots \left(1.2\frac{sta}{ft}\right)(2\ sta) + 1232.98\ ft = 1234.39\ ft$$

Problem

A vertical sag curve has the following properties:
Curve length (horizontal projection), L = 500 feet
Grade from vertex to EVC, G_2 = 1.85%
Rate of grade change, R = 0.87% per station

Calculate the distance (in feet) from the EVC to the turning point.

Solution
The horizontal projection of the curve length, L, is equal to 500 ft/(100 ft/sta) = 5 stations.
Calculate the grade, G_1, from the BVC to the vertex from the following equation:
$$R = \frac{G_2 - G_1}{L}; G_1 = G_2 - RL$$
$$G_1 = 1.85\% - \left(\frac{0.87\%}{station}\right)(5\ stations) = -2.5\%$$
Next, calculate the distance in stations, x, from the BVC to the turning point using the following equation:
$$x_{BVC\ to\ TP} = \frac{-G_1}{R} = \frac{-(-2.5\%)}{0.87\%/station} = 2.87\ stations$$
Since the total distance from the BVC to the EVC is 5 stations (500 feet), the distance from the turning point to the EVC is:
$$x_{TP\ to\ EVC} = 5\ stations - 2.87\ stations = 2.13\ stations$$
$$x_{TP\ to\ EVC} = 2.13\ stations\left(\frac{100\ ft}{1\ station}\right) = 213\ feet$$

Problem

A vertical sag curve's vertex and turning point have elevations of 1890 feet and 1750 feet, respectively.
Calculate the curve length if the curve has the following properties:
Grade from BVC to vertex = -2.30%
Grade from vertex to EVC = 4.20%

Solution
For situations in which the vertical curve needs to pass through the turning point at a given elevation, the following equation can be used to directly calculate the vertical curve length, L, that would result from this requirement:
$$L = \frac{2(Elev_V - Elev_{TP})}{G_1\left(\frac{G_1}{G_2 - G_1} + 1\right)}$$
Where $Elev_V$ is the elevation of the vertex, $Elev_{TP}$ is the elevation of the turning point, G_1 is the grade from the BVC to the vertex, and G_2 is the grade from the vertex to the EVC.
$$L = \frac{2(1890\ ft - 1750\ ft)}{-2.30\ ft/sta\left(\frac{-2.30\ ft/sta}{4.20\ ft/sta - (-2.30 ft/sta)} + 1\right)}$$
$$L = 89.92\ feet$$

Problem

Calculate the required intersection sight distance along a major road for a vehicle traveling along a minor road that has to yield to cross a major road (AASHTO Case C1) given the following information:

Design vehicle length = 34 feet
Intersection Width = 40 feet
Minor Road Design Speed = 30 mph
Major Road Design Speed = 45 mph
Travel Time to Reach Major Road from Decision Point = 3 seconds

Solution
Per AASHTO's 2004 design guidelines, the time gap, t_g, in seconds for Case C1 control is calculated using the following equation:
$$t_g = t_a + \frac{w + L_a}{0.88 V_{minor}}$$
Where t_a is the travel time to reach the major road from the driver decision point in seconds, w is the intersection width in feet, L_a is the design vehicle length in feet, and V_{minor} is the design speed of the minor road in miles per hour.
$$t_g = 3\,s + \frac{40\,ft + 34\,ft}{0.88(30\,mph)} = 5.8\,s$$
The intersection sight distance, *ISD*, in feet can then be calculated from:
$$ISD = 1.47 V_{major} t_g$$
Where V_{major} is the design speed of the major road in miles per hour.
$$ISD = 1.47(45\,mph)(5.8\,s) = 384\,ft$$

Problem

Calculate the minimum curve length for stopping sight distance of a vertical crest curve for the following conditions:
Algebraic difference in the vertical crest curve grades = 5%
Height of the driver's eyes in feet = 3.0 ft
Height of object sighted = 2.0 ft
Required stopping sight distance = 250 ft

Solution
There are two equations provided in AASHTO's 2004 design guidelines for the calculation of the required curve length, L (in feet), based on its value in comparison to the required stopping sight distance, S (in feet). Since L is unknown, both values of L are calculated to determine which is valid.

For $S < L$, $$L = \frac{AS^2}{200\left(\sqrt{h_1} + \sqrt{h_2}\right)^2}$$

For $S > L$, $$L = 2S - \frac{200\left(\sqrt{h_1} + \sqrt{h_2}\right)^2}{A}$$

Where A is the difference in the vertical curve grades, in percent, h_1 is the height of the driver's eyes in feet, and h_2 is the height of the object sighted.

$$S < L, \quad L = \frac{(5\%)(250\ ft)^2}{200(\sqrt{3\ ft} + \sqrt{2\ ft})^2} = 157.84\ ft$$

$$S > L, \quad L = 2(250\ ft) - \frac{200(\sqrt{3\ ft} + \sqrt{2\ ft})^2}{5\%} = 104.04\ ft$$

The latter value meets the condition, so L = 104.04 ft.

Problem

Calculate the minimum curve length for stopping sight distance of a vertical sag curve for the following conditions:
Algebraic difference in the vertical sag curve grades = 3.75%
Height of the driver's eyes in feet = 3.5 ft
Height of object sighted = 2.0 ft
Required stopping sight distance, S = 460 ft

Solution
For the given heights of the driver's eyes and the object sighted, there are two equations provided in AASHTO's 2004 design guidelines for the calculation of the required curve length, L (in feet) for stopping sight distance. Since L is unknown, both values of L are calculated to determine which condition is valid.

For $S < L$, $\quad L = \dfrac{AS^2}{400 + 3.5S}$

For $S > L$, $\quad L = 2S - \dfrac{400 + 3.5S}{A}$

Where A is the difference in the vertical curve grades, in percent.

For $S < L$, $\quad L = \dfrac{(3.75\%)(460\ ft)^2}{400 + 3.5(460\ ft)} = 394.78\ ft$

For $S > L$, $\quad L = 2(460\ ft) - \dfrac{400 + 3.5(460\ ft)}{3.75\%} = 384.0\ ft$

The latter value meets the condition, so L = 384.0 ft.

Problem

Calculate the minimum curve length for passing sight distance of a vertical crest curve for the following conditions:
Algebraic difference in the vertical crest curve grades = 4.25%
Height of the driver's eyes in feet = 3.5 ft
Height of object sighted = 3.5 ft
Required passing sight distance, S = 475 ft

Solution
For the given heights of the driver's eyes and the object sighted, there are two equations provided in AASHTO's 2004 design guidelines for the calculation of the required curve length, L (in feet) for passing sight distance. Since L is unknown, both values of L are calculated to determine which condition is valid.

For $S < L$, $\quad L = \dfrac{AS^2}{2800}$

$$\text{For } S > L, \quad L = 2S - \frac{2800}{A}$$
Where A is the difference in the vertical curve grades, in percent.
$$\text{For } S < L, \quad L = \frac{(4.25\%)(475\ ft)^2}{2800} = 342.47\ ft$$
$$\text{For } S > L, \quad L = 2(475\ ft) - \frac{2800}{4.25\%} = 291.18\ ft$$
The latter value meets the condition, so L = 291.2 ft.

Problem

Calculate the minimum curve length for stopping sight distance of a vertical sag curve with an overhead structure for the following conditions:
Algebraic difference in the vertical sag curve grades = 22.0%
Height of the driver's eyes in feet, h_1 = 3.5 ft
Height of object sighted, h_2 = 6.5 ft
Required stopping sight distance, S = 490 ft
Vertical clearance in feet for overhead structure, C = 17 ft

Solution
Per AASHTO's 2004 design guidelines for the calculation of the required curve length, L (in feet) for stopping sight distance obstructed by an overhead structure:
$$\text{If } S < L, \quad L = \left(\frac{AS^2}{800}\right)\left(C - \frac{h_1 + h_2}{2}\right)^{-1}$$
$$\text{If } S > L, \quad L = 2S - \left(\frac{800}{A}\right)\left(C - \frac{h_1 + h_2}{2}\right)$$
Where A is the difference in the vertical curve grades, in percent.
$$S < L, \quad L = \left(\frac{(22.0\%)(490\ ft)^2}{800}\right)\left(17\ ft - \frac{3.5\ ft + 6.5\ ft}{2}\right)^{-1}$$
$$S < L, \quad L = 550.23\ ft$$
$$S > L, \quad L = 2(490\ ft) - \left(\frac{800}{(22.0\%)}\right)\left(17\ ft - \frac{3.5\ ft + 6.5\ ft}{2}\right)$$
$$S > L, \quad L = 543.64\ ft$$
The former value meets the condition, so L = 550.23 ft.

Problem

Calculate the stopping sight distances for a vehicle traveling at 30 miles per hour on a straight paved roadway having a constant uphill grade of 3%. Assume the braking perception-reaction time is 2.5 seconds and the coefficient of friction for the wet pavement is 0.35. Also calculate the stopping sight distance if the constant grade was 3% downhill.

Solution
For a straight roadway with a constant grade, the stopping sight distance, S, may be calculated from:
$$S = \left(\frac{1.47\ ft/s}{mi/hr}\right)t_p v_{mph} + \frac{v_{mph}^2}{30\ (f + G)}$$

Where t_p is the braking perception-reaction time, v_{mph} is the vehicle speed, f is the coefficient of friction, and G is the roadway grade. The stopping sight distance is then:

$$S = \left(\frac{1.47 \frac{ft}{s}}{\frac{mi}{hr}}\right)(2.5\ s)(30\ mph) + \frac{(30\ mph)^2}{30\ (0.35 + 0.03)}$$

$S = 189.20\ ft\ (uphill)$

For a downhill grade, the only thing that changes is the sign of G:

$$S = \left(\frac{1.47 \frac{ft}{s}}{\frac{mi}{hr}}\right)(2.5\ s)(30\ mph) + \frac{(30\ mph)^2}{30\ (0.35 - 0.03)}$$

$$S = 204.0\ ft\ (downhill)$$

Problem

Show the apparent centrifugal force and vehicle weight force components for a vehicle traveling on a superelevated horizontal curve.

Solution

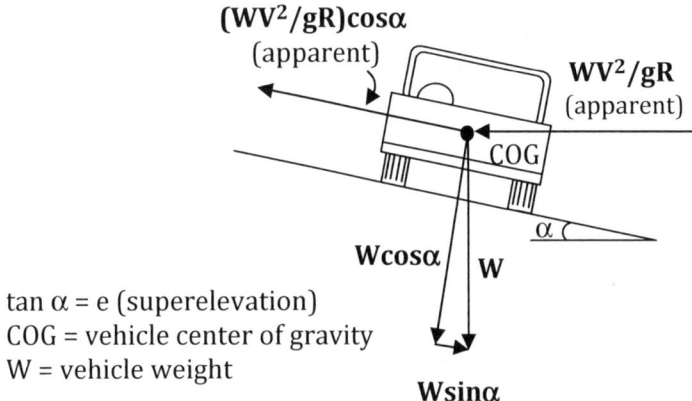

tan α = e (superelevation)
COG = vehicle center of gravity
W = vehicle weight

Problem

A horizontal curve has a radius of 1340 feet. Accounting for sideways friction, calculate the superelevation (in degrees) if the vehicle speed is 65 miles per hour.

Solution

For vehicular speeds 50 to 70 miles per hour, the following equation can be used to calculate the side friction factor, f_s:

$$f_s = 0.14 - \frac{0.02(v_{mph} - 50)}{10}$$

Where v_{mph} is the vehicular velocity in miles per hour.

$$f_s = 0.14 - \frac{0.02(65 - 50)}{10} = 0.11$$

The superelevation, e, as a ratio can then be calculated from:

$$e = \frac{v_{mph}^2}{15R} - f_s$$

Where R is the instantaneous curve radius in feet.
$$e = \frac{(65\ mph)^2}{15(1340\ ft)} - 0.11 = 0.10$$
The superelevation, in degrees, is:
$$e = \tan \alpha;\ \alpha = \tan^{-1} e = \tan^{-1} 0.10$$
$$\alpha = 5.71°$$

Problem

A horizontal curve has a radius of 600 feet. Accounting for sideways friction, calculate the centrifugal factor and the superelevation (in degrees) if the vehicle speed is 45 miles per hour.

Solution
For vehicular speeds less than 50 miles per hour, the following equation can be used to calculate the side friction factor, f_s:
$$f_s = 0.16 - \frac{0.01(v_{mph} - 30)}{10}$$
Where v_{mph} is the vehicular velocity in miles per hour.
$$f_s = 0.16 - \frac{0.01(45 - 30)}{10} = 0.145$$
Centrifugal factor is defined as $e + f_s$ and is calculated from:
$$e + f_s = \frac{v_{mph}^2}{15R}$$
Where R is the instantaneous curve radius in feet.
$$e + f_s = \frac{(45\ mph)^2}{15(600\ ft)} = 0.225$$
The superelevation is then:
$$e = 0.225 - f_s = 0.225 - 0.145 = 0.08$$
The superelevation, in degrees, is:
$$e = \tan \alpha;\ \alpha = \tan^{-1} e = \tan^{-1} 0.08$$
$$\alpha = 4.57°$$

Terms

Intentional banking of roadways may be included in the design to offset the centrifugal forces resulting from a vehicle traveling on a horizontal curve. If a roadway is banked, the difference in elevations of the outside and inside curve is called the *superelevation*.

Tangent runout (also referred to as the tangent runoff) is the distance from which the roadway section is normally crowned to the location that the crowned slope of the outer curve (adverse crown) has been removed. The end point of the tangent runout is the starting point of the superelevation runoff.

Superelevation runoff is the distance from which a roadway section has had the adverse crown removed to the location that the roadway is completely superelevated (banked to offset centrifugal forces).
The *superelevation transition distance* is the sum of the tangent runout and the superelevation runoff distances.

Problem

Calculate the tangent runout and the superelevation runoff distances for a 12-ft lane with a superelevation runoff rate equal to 1:300, a required superelevation of 0.05, and a cross slope rate of 0.015 ft/ft.

Solution
The tangent runout (or tangent runoff) can be calculated from the following:
$$T_R = \frac{w \times p}{SRR}$$
Where T_R is the tangent runout, w is the lane width, p is the cross slope rate, and SRR is the superelevation runoff rate. T_R is then:
$$T_R = \frac{12\ ft \times 0.015\ ft/ft}{\frac{1\ ft}{300\ ft}} = 54\ ft$$
The superelevation runoff distance can be calculated from the following equation:
$$L = \frac{w \times e}{SRR}$$
Where L is the superelevation runoff distance, and e is the superelevation. L is then:
$$L = \frac{12\ ft \times 0.05\ ft/ft}{\frac{1\ ft}{300\ ft}} = 180\ ft$$

Problem

A horizontal curve on a highway as the following properties:
Superelevation runoff distance, L = 180 ft
Tangent runout distance, T_R = 54 ft
Stationing at point of curvature = 35+00

Calculate the stationing at which a) the superelevation runoff, L, begins, b) the tangent runout, T_R, begins, and c) the pavement is completely superelevated.

Solution
A rule of thumb typically used for state highways is that two-thirds of the superelevation runoff distance is located on the horizontal curve tangent, and the remaining one-third is located on the curve itself. The point at which the superelevation runoff begins is then located at two-thirds of L before the PC:
$$(sta\ 35 + 00) - \frac{2}{3}(180\ ft) = sta\ 33 + 80$$

b) The beginning of the superelevation runoff is also the ending of the tangent runout. The point at which the tangent runout begins is equal to the tangent runout distance subtracted from the superelevation runoff begin point:
$$(sta\ 33 + 80) - 54\ ft = sta\ 33 + 26$$

c) The point at which the pavement is completely superelevated is located at the end of the superelevation runoff distance, which will be equal to one-third of L after the PC stationing:
$$(sta\ 35 + 00) + \frac{1}{3}(180\ ft) = sta\ 35 + 60$$

Problem

Define PIEV and calculate the PIEV for a vehicle traveling at 50 miles per hour if the perception-reaction distance is 169 feet.

Solution
PIEV is the time it takes the following to occur after a driver of a vehicle is affected by a particular stimulus: perception, identification, emotion (decision), and volition (reaction). It is also referred to as the perception-reaction time and is important for designing roadways and traffic light timing. The PIEV does not take into account the time it takes for the vehicle to stop after the brakes have been applied. The PIEV is calculated from the following equation:

$$PIEV = \frac{d_p}{1.47 \times v}$$

Where d_p is the perception-reaction distance (the distance the vehicle travels while the PIEV steps occur), v is the vehicle velocity in miles per hour (mph) and 1.47 is the conversion factor from mph to feet per second. The PIEV is then:

$$PIEV = \frac{169 \text{ feet}}{1.47 \times (50 \text{ mph})} = 2.3 \text{ seconds}$$

Problem

A driver of a vehicle traveling at 70 miles per hour has a PIEV time of 2.0 seconds. Determine if the driver will stop before hitting an obstruction located 450 feet ahead of him if the brakes are applied and the vehicle is slowed at a constant deceleration of 18 ft/s².

Solution
First calculate the distance traveled during the driver PIEV:
$d_p = PIEV(1.47 \times v_o) = (2.0 \text{ s})(1.47 \times 70 \text{ mph}) = 205.8 \text{ ft}$
The time it will take for the vehicle to come to a complete stop after the brakes are applied is equal to $(v-v_o)/a$, where v is the final vehicle velocity, v_o is the initial vehicle velocity, and a is the constant acceleration or deceleration rate (negative if deceleration). The time is then:

$$t = \frac{(0 - 70 \text{ mph})(1.47 \frac{ft}{s}/mph)}{-18 \text{ ft/s}^2} = 5.72 \text{ s}$$

The distance traveled over 5.72 seconds can be calculated from:
$$d = v_o t + \frac{1}{2}at^2$$
$$= (70 \text{ mph})(1.47 \frac{ft}{s}/mph)(5.72s) + \frac{1}{2}\left(-18\frac{ft}{s^2}\right)(5.72 \text{ s})^2$$
$$= 294 \text{ ft}$$

The total stopping distance including PIEV is 205.8 ft + 294.1 ft = 499.9 ft. The driver would not be able to stop in time.

Problem

Determine the safe stopping distance for a vehicle traveling at 65 mph uphill at a grade of 2% that slows at a constant deceleration of 11.2 ft/s² after the brakes have been applied. Assume the driver PIEV time is 2.5 seconds.

Solution
The braking distance, taking grade into account, is calculated from:
$$d_b = \frac{v_o^2 - v^2}{30\left[\left(\frac{a}{32.2}\right) \pm G\right]}$$
Where d_b is the braking distance, v_o is the initial vehicle velocity, v is the final vehicle velocity, a is the constant deceleration rate, and G is the roadway grade (negative for downhill and positive for uphill). The braking distance is then:
$$d_b = \frac{(65\ mph)^2 - 0^2}{30\left[\left(\frac{11.2\frac{ft}{s^2}}{32.2}\right) + 0.02\right]} = 382.9\ ft$$
The distance traveled during the driver PIEV:
$d_p = PIEV(1.47 \times v_o) = (2.5\ s)(1.47 \times 65\ mph) = 238.9\ ft$
The safe stopping distance, d_s, is then:
$d_s = d_p + d_b = 238.9\ ft + 382.9\ ft = 621.8\ ft$

Problem

Provide the equations for calculating the minimum passing sight distance.

Solution
Per AASHTO (2004), the minimum passing sight distance is the summation of $d_1 + d_2 + d_3 + d_4$, where d_1 is the distance traveled during the PIEV driver time and the initial acceleration to the point the driver is going to cross into the left lane for passing:
$$d_1 = 1.47 t_i \left[v - m + \frac{1}{2}(at_i)\right]$$
Where t_i is the time of the initial maneuver, m is the difference of the speed between the passing and passed vehicles, v is the average speed of the passing vehicle, and a is the average acceleration of the passing vehicle.
The variable d_2 is the distance traveled while the passing vehicle is in the passing lane: $d_2 = 1.47vt_2$, where t_2 is the time the passing vehicle is in the left lane. The distance between the passing vehicle and the opposing vehicle at the end of the passing maneuver is known as d_3. Finally, d_4 is the distance traveled by the opposing vehicle for 2/3 of the time the passing vehicle is in the left lane: $d_4 = 2/3 d_2$.

Problem

List the horizontal and vertical clearances per AASHTO (2004) guidelines, as available, for local rural roads, local urban roads, rural collectors, urban collectors, rural and divided arterials, urban arterials, and freeways.

Solution
For local rural roads, the vertical clearance is 14 feet, and the horizontal clearance is 7-10 feet. For local urban roads, the vertical clearance is 14 feet, and the horizontal clearance is greater than or equal to 1.5 feet. For rural collectors, the vertical clearance is 14 feet, and the horizontal clearance is 10 feet minimum. For urban collectors, the vertical clearance is 14 feet plus an additional allowance for future resurfacing. The horizontal clearance for

urban collectors is greater than or equal to 1.5 feet from the curb. The vertical clearance for rural and divided arterials is 16 feet. The vertical distance for urban arterials is also 16 feet. For freeways in general, the vertical clearance is greater than or equal to 16 feet, (17 feet for signs). The horizontal clearance for freeways is per AASHTO's clear zone concept.

Problem

The driver of a 3500-lb (mass) vehicle traveling at 75 miles per hour hits the brakes and decelerates at a constant rate for 8.5 seconds before stopping. Calculate the frictional (retarding) force and the coefficient of friction between the vehicle tires and the road.

Solution
First, the acceleration or deceleration rate, a, is calculated from the following equation for uniform acceleration or deceleration:

$$a = \frac{v - v_o}{t} = \frac{(0 - 75\ mph)(1.47 \frac{ft}{s}/mph)}{8.5\ s} = -13.0\ ft/s^2$$

This deceleration is then used in the following equation to calculate the frictional (retarding) force:

$$F_{frictional} = ma = \left(\frac{3500\ lbm}{32.2\ \frac{ft \times lbm}{lbf \times s^2}}\right)\left(13.0 \frac{ft}{s^2}\right) = 1413\ lbf$$

The coefficient of friction, f, between the vehicle tires and the road is then calculated from:

$$f = \frac{F_{frictional}}{N}$$

Where N is the normal force (weight) of the vehicle.

$$f = \frac{1413\ lbf}{3500\ lbf} = 0.40$$

Terms

Average annual daily traffic (AADT) is the average 24-hour traffic volume for a particular location over a full year. It is calculated by dividing the total annual traffic by 365 days. The *average annual weekday traffic (AAWT)* is the average 24-hour traffic volume for a particular location on weekdays only over a full year. The *average daily traffic (ADT)* is an estimate of 24-hour traffic volume for a particular location over time periods less than a full year. The *average weekday traffic (AWT)* is an estimate of the 24-hour traffic volume for a particular location on weekdays only over time periods less than a full year. The averages and estimations may be for one lane, for all lanes in one direction, or for all lanes in both directions. The *peak hour factor (PHF)* is the ratio of the total actual hourly traffic volume to the peak rate of traffic flow within the hour.

Problem

The information recorded during a traffic study at a particular location is displayed below. Determine what the peak hourly traffic volume is from this information.

Time of Day	Volume (No. of Vehicles)
7:00 – 7:15 am	140
7:15 – 7:30 am	178
7:30 – 7:45 am	230
7:45 – 8:00 am	265
8:00 – 8:15 am	233
8:15 – 8:30 am	185
8:30 – 8:45 am	215
8:45 – 9:00 am	240
9:00 – 9:15 am	210
9:15 – 9:30 am	170

Solution

To calculate the peak hourly traffic volume from the recorded information, the hourly totals for each available interval must be calculated. The hourly traffic from the 7:00 am to 8:00 time interval is: $7:00 - 8:00 am$: $140 + 178 + 230 + 265 = 813$

The remaining hourly volumes are calculated similarly for each hour that data is available:

$7:15 - 8:15 am$: $178 + 230 + 265 + 233 = 906$
$7:30 - 8:30 am$: $230 + 265 + 233 + 185 = 913$
$7:45 - 8:45 am$: $265 + 233 + 185 + 215 = 898$
$8:00 - 9:00 am$: $233 + 185 + 215 + 240 = 873$
$8:15 - 9:15 am$: $185 + 215 + 240 + 210 = 850$
$8:30 - 9:30 am$: $215 + 240 + 210 + 170 = 835$

The peak hourly traffic volume is thus 913 vehicles per hour (which was evidenced between 7:30 am and 8:30 am).

Problem

The information recorded during a traffic study at a particular location is below. The peak hourly traffic volume from this information is determined to be 913 vehicles per hour (between 7:30 am and 8:30 am). Calculate the peak hour factor.

Time of Day	Volume (No. of Vehicles)
7:00 – 7:15 am	140
7:15 – 7:30 am	178
7:30 – 7:45 am	230
7:45 – 8:00 am	265
8:00 – 8:15 am	233
8:15 – 8:30 am	185
8:30 – 8:45 am	215
8:45 – 9:00 am	240
9:00 – 9:15 am	210
9:15 – 9:30 am	170

Solution
The peak hour factor, PHF, is the ratio of the total actual hourly traffic volume to the peak rate of traffic flow within the hour as seen in the equation below:

$$PHF = \frac{V_{vph}}{v_p}$$

Where V_{vph} is the actual peak hourly traffic volume (913 vehicles per hour for this problem) and v_p is the peak rate of traffic flow within the hour. The peak rate of traffic flow within the hour is determined by finding the interval with the greatest traffic volume and multiplying that by the appropriate number to obtain an hourly peak value. The peak volume from the data given is 265 vehicles in the 15 minute interval of 7:45 – 8:00 am. The peak rate of traffic flow within an hour is then (265 vehicles/15 min interval) x (4-15 min interval/1 hour) = 1060 vehicles per hour. The PHF is then:

$$PHF = \frac{913 \ vehicles/hour}{1060 \ vehicles/hour} = 0.861$$

Problem

A 3750 lbm vehicle travels around a curve at a speed of 55 miles per hour. Calculate the centripetal force for curve radii of 400 feet and 600 feet.

Solution
The centripetal force is calculated from the following equation:

$$F_c = \frac{mv_t^2}{g_c * r}$$

Where *m* is the vehicle mass, v_t is the tangential velocity, g_c is the gravitational constant, and *r* is the curve radius.

For a curve radius of 400 feet, the equation becomes:

$$F_c = \frac{(3750 \ lbm)\left[(55 \ mph)\left(\frac{1.47 \frac{ft}{s}}{mph}\right)\right]^2}{\left(\frac{32.2 \ lbm \times ft}{lbf \times s^2}\right) 400 \ ft} = 1903 \ lbf$$

For a curve radius of 600 feet, the equation becomes:

$$F_c = \frac{(3750 \ lbm)\left[(55 \ mph)\left(\frac{1.47 \frac{ft}{s}}{mph}\right)\right]^2}{\left(\frac{32.2 \ lbm \times ft}{lbf \times s^2}\right) 600 \ ft} = 1269 \ lbf$$

Problem

A vehicle travels along a horizontal curve (radius of 750 feet) at a speed of 50 miles per hour. Calculate the superelevation (in degrees) required for the vehicle to offset the centrifugal forces without having to rely on tire friction.

Solution
The superelevation, e, as a ratio is calculated from:
$$e = \frac{v_{mph}^2}{15R} - f_s$$
Where v_{mph} is the vehicular velocity in miles per hour, R is the instantaneous curve radius (in feet), and f_s is the side friction factor. If the banking or superelevation will offset the centrifugal forces, the side friction factor will be equal to zero. The equation then becomes:
$$e = \frac{v_{mph}^2}{15R} - 0 = \frac{v_{mph}^2}{15R}$$
$$e = \frac{(50\ mph)^2}{15(750\ ft)} = 0.22$$
The superelevation, in degrees, is:
$e = \tan \alpha;\ \alpha = \tan^{-1} e = \tan^{-1} 0.22$
$\alpha = 12.41°$

Problem

The time it takes each of six vehicles to travel 1600 meters is recorded and shown below. Calculate the time mean speed and the space mean speed for the provided data.
Vehicle 1: 49 seconds
Vehicle 2: 51 seconds
Vehicle 3: 55 seconds
Vehicle 4: 45 seconds
Vehicle 5: 53 seconds

Solution
The time mean speed is the average speed it takes the vehicles to pass a common point on the roadway over a given time period. The equation for calculating the time mean speed, μ_t, is:
$$\mu_t = \frac{\sum \left(\frac{d}{t_i}\right)}{n}$$
Where d is the distance traveled, t_i is the travel time of the *i*th vehicle, and n is the total number of travel times observed.

The time mean speed is then:
$$\mu_t = \frac{\frac{1600\ m}{49\ s} + \frac{1600\ m}{51\ s} + \frac{1600\ m}{55\ s} + \frac{1600\ m}{45\ s} + \frac{1600\ m}{53\ s}}{5}$$
$\mu_t = 31.8\ m/s$

The space mean speed, μ_s, is the average speed by means of calculating the instantaneous speeds over a roadway section:
$$\mu_s = \frac{nd}{\sum t_i} = \frac{5(1600\ m)}{49\ s + 51s + 55s + 45s + 53s} = 31.6\ m/s$$

Problem

Determine the average vehicle speed for a single lane roadway having an average headway of 2.8 seconds per vehicle and an average vehicle spacing of 180 feet.

First, the vehicle flow rate is calculated. The rate of flow, or number of vehicles crossing a point per hour per lane, is equal to the reciprocal of the headway, the time between successive vehicles: $q = 1/h$, where q is the vehicle flow rate, and h is the headway in seconds per vehicle.

$$q = \frac{1}{2.8 \, s/vehicle} \left(\frac{3600 \, seconds}{1 \, hour} \right) = 1286 \, veh/hr$$

The average speed of the traffic, u, can be calculated from: $u = q/k$, where q is in units of vehicles per hour, and k is the density of the traffic in units of vehicles per unit distance. The density of traffic is the inverse of the vehicle spacing:

$$k = \left(\frac{1}{\frac{180 \, ft}{vehicle}} \right) \left(\frac{5280 \, ft}{1 \, mile} \right) = 29.3 \, vehicles/mile$$

$$u = \frac{1286 \, vehicles/hour}{29.3 \, vehicles/mile} = 43.9 \, miles \, per \, hour$$

Spiral curve

A spiral curve is a curve that provides a more gradual transition from the tangent of a circular curve to the circular curve itself. A spiral curve does not have a constant radius; the radius of the spiral curve at the end of the tangent is considered infinity and uniformly reduces to match the radius of the circular curve at the circular curve's point of tangency. For a curve system that has matching spiral curves, the total length of the curve system, L, is equal to the sum of the length of the circular curve, $L_{circular}$, and both spiral curve lengths, $2 \times L_{spiral}$:

$$L = l_{circular} + 2L_{spiral}$$

For spiral curves, the following relationship is used in calculations:

$$\frac{R_{spiral}}{R_{circular}} = \frac{L_{spiral}}{L_{circular}}$$

Where R_{spiral} is the radius of the spiral curve, $R_{circular}$ is the radius of the circular curve, L_{spiral} is the length of the spiral curve, and $L_{circular}$ is the length of the circular curve.

Water Resources and Environmental

Properties of liquid fluids

Fluid Shape: The boundaries of a fluid are limited and defined by the container holding the fluid. The fluid's volume remains the same, regardless of the fluid's container, and does not substantially change as a result of pressure and temperature changes.
Fluids under Pressure and Compressibility: If a fluid is under pressure, the pressure at any point within the fluid is equal in all directions. Fluids may compress slightly under high pressures, but for most applications, fluids are typically considered incompressible.
Relationship between Density and Temperature: As the temperature of a liquid increases, the density of the fluid will decrease.
Real vs. Ideal Fluids: Real fluids a) are compressible, b) have limited viscosities, c) are subjected to friction and turbulence during flow, and d) exhibit non-uniform velocity distributions. Fluids that are considered ideal a) are incompressible, b) have no viscosity, c) do not experience friction and turbulence during flow, and d) exhibit uniform velocity distributions.

Problem

Explain gage pressure, absolute pressure, and vacuum pressure. Also calculate the absolute pressure for a gage pressure of 0.752 bar (reading taken at Standard Temperature and Pressure air conditions).

Solution
Gage pressure refers to a pressure that is measured with respect to atmospheric pressures. Absolute pressure refers to a pressure that is measured with respect to true zero pressure. Vacuum pressure refers to a pressure below atmospheric pressure. The following equations show how gage pressure, absolute pressure, atmospheric pressure, and vacuum pressure are related to one another:

$p_{absolute} = p_{gage} + p_{atmosphere}$

$p_{absolute} = p_{atmospheric} - p_{vacuum}$

Where $p_{atmosphere}$ is the actual pressure of the atmosphere that is measured at the same time the gage pressure measurement is taken. Air at Standard Temperature and Pressure conditions is 0°C and is at sea level with a pressure value of 1.000 atm (1.01325 bar). For a gage pressure of 0.752 bar and a $p_{atmosphere}$ of 1.01325 bar, the absolute pressure is:
$$p_{absolute} = 0.752\ bar + 1.01325\ bar = 1.765\ bar$$

Problem

Calculate the specific gravity of oxygen gas at a temperature of 351 Kelvin and a pressure of 1.35×10^5 Pa. Use air at Standard Temperature and Pressure (STP) conditions as the reference.

Solution
First calculate the specific gas constant, $R_{specific}$ for oxygen gas:

$$R_{specific} = \frac{R}{MW}$$

Where R is the universal gas constant (8.314 J/mol*K) and MW is the molecular weight of oxygen gas (2 × 16 = 32 grams per mole).

$$R_{specific} = \frac{8.314 \frac{J}{mol \times K}}{32 \, g/mol} = 0.26 \frac{J}{g \times K} = 259.8 \frac{J}{kg \times K}$$

Using the ideal gas law to calculate the density, ρ:

$$\rho = \frac{p}{R_{specific}T} = \frac{1.35 \times 10^5 \, Pa \left(\frac{1 \frac{J}{m^3}}{1 \, Pa}\right)}{259.8 \frac{J}{kg \times K}(351 \, K)} = 1.48 \, kg/m^3$$

The specific gravity, SG, is calculated by dividing the density of the oxygen gas by the density of air at STP (equal to 1.29 kg/m³):

$$SG = \frac{1.48 \, kg/m^3}{1.29 \, kg/m^3} = 1.15$$

Osmosis and reverse osmosis

Consider two solutions having unequal concentrations separated by a semi-permeable barrier that is only semi-permeable to the solvent and not the solute. *Osmosis* is the passage of the solvent from the side having the lower concentration through the semi-permeable barrier to the side with the higher concentration. This will continue (and the water level of the side with the higher concentration will rise) until the concentrations on both sides of the barrier become equal. This diffusion of water across the semi-permeable barrier creates what is called osmotic pressure. If a pressure equaling the osmotic pressure is applied to the solution having the higher concentration, diffusion of the solvent will cease. If the pressure applied to the solution having the higher concentration exceeds the osmotic pressure, the solvent will be forced back to the side with the lower concentration, creating one side with a substantially higher concentration than the other. This process is called *reverse osmosis*, and is a used all over the world for a variety of water treatment applications.

Energy conservation equation and Bernoulli energy conservation equation

The *energy conservation equation* assumes the flow between points 1 and 2 is one-dimensional, steady, and incompressible:

$$E_{v,1} + E_{p,1} + E_{z,1} = E_{v,2} + E_{p,2} + E_{z,2} + h_L + E_m$$

Where $E_{v,1}$ and $E_{v,2}$ are the velocity heads or kinetic energies at points 1 and 2, respectively, $E_{p,1}$ and $E_{p,2}$ are the pressure heads or pressure energies at points 1 and 2, respectively, and $E_{z,1}$ and $E_{z,2}$ are the elevation heads or potential energies at points 1 and 2, respectively. The component h_L is the head loss between points 1 and 2, and E_m is any mechanical energy added to the system. The velocity, pressure and elevations heads in the equation can be rewritten as:

$$\frac{v_1^2}{2g_c} + \frac{p_1}{\rho} + \frac{z_1 g}{g_c} = \frac{v_2^2}{2g_c} + \frac{p_2}{\rho} + \frac{z_2 g}{g_c} + h_L + E_m$$

For ideal fluid flow in which the system experiences no energy losses or gains, the classic *Bernoulli energy conservation equation* is derived from the energy conservation equation:

$$\frac{v_1^2}{2g_c} + \frac{p_1}{\rho} + \frac{z_1 g}{g_c} = \frac{v_2^2}{2g_c} + \frac{p_2}{\rho} + \frac{z_2 g}{g_c}$$

Continuity equation

The *continuity equation* states that fluid mass is conserved in fluid systems. The equation for mass conservation is:
$$\dot{m}_1 = \dot{m}_2$$
Where $\dot{m}$ is the mass flow rate. For fluid flow, the continuity equation between points 1 and 2 is:
$$\rho_1 A_1 v_1 = \rho_2 A_2 v_2$$
Where ρ is the fluid density, A is the cross sectional area with respect to the flowing fluid, and v is the flow velocity. If the fluid is incompressible, the densities of the fluid at point 1 and point 2 are equal:
$$\rho_1 = \rho_2$$
And the continuity equation becomes:
$$A_1 v_1 = A_2 v_2$$
The term Av is also referred to as the volumetric flow rate, Q. So the volumetric flow rate at point 1 is equal to the volumetric flow rate at point 2:
$$Q_1 = Q_2$$

Problem

A vertical circular plane area of radius 2 feet is submerged in water. The centroid of the circular plane is located 20 feet below the water surface. Calculate the magnitude of the hydrostatic force on the submerged surface.

Solution
The hydrostatic force, F_H, on a submerged plane surface is calculated from:
$$F_H = \gamma h_c A$$
Where γ is the specific weight of the fluid, h_c is the submerged depth of the plane area centroid, and A is the surface area of the plane being acted on.
The surface area of the plane being acted on in this example is:
$$A = \pi r^2 = \pi (2\ ft)^2 = 12.6\ ft^2$$
The submerged depth of the circular plane area centroid is 20 feet (given). The specific weight of water is 62.4 lbf/ft³.
The magnitude of the hydrostatic force acting on the submerged circular plane area is then:
$$F_H = \left(62.4 \frac{lbf}{ft^3}\right)(20\ ft)(12.6\ ft^2) = 15{,}725\ lbf$$

Problem

A sedimentation tank with straight sides (case a) and inclined sides (case b) is shown below. Discuss the differences in the depth to the resultant hydrostatic force and the average pressure between the two cases for the plane surface located between h_2 and h_1.

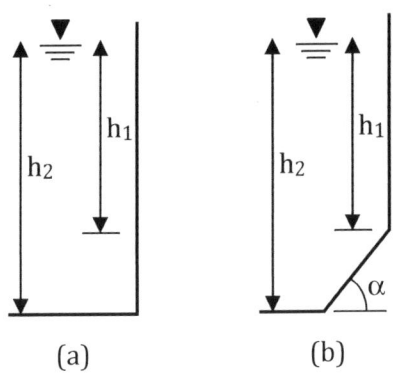

(a) (b)

Solution
For case (a), the depth from the water surface to the resultant hydrostatic force, H_{HF}, is calculated from the following equation:

$$H_{HF} = \frac{2}{3}\left(h_1 + h_2 - \frac{h_1 h_2}{h_1 + h_2}\right)$$

For case (b), the angle of the inclination must be taken into account to determine the depth from the water surface to the resultant hydrostatic force. The equation for calculating the location of this force on an inclined surface is shown below:

$$H_{HF} = \frac{\frac{2}{3}\left(h_1 + h_2 - \frac{h_1 h_2}{h_1 + h_2}\right)}{\sin\theta}$$

The average pressure, $\bar{p}$, from which the resultant hydrostatic force is calculated, does not change between cases (a) and (b). The average pressure is calculated from the following equation:

$$\bar{p} = \frac{1}{2}\gamma(h_1 + h_2)$$

Where γ is the specific weight of the fluid.

Problem

A pipe exits from the bottom of a side of an open-air rectangular water tank. The free water surface elevation of the water in the tank is 2500 feet. Water flows freely from the pipe discharge point at an elevation of 2250 feet. Calculate the total energy of the system and the velocity of the water at the pipe discharge. Assume there is no fluid friction, the fluid is incompressible, and changes in thermal energy are negligible.

Solution
Based on the assumptions, the Bernoulli equation may be used to calculate the total energy. The total energy for the system will consist of kinetic, pressure, and potential energy components. At the tank water surface, there is no kinetic or pressure energy. The only energy present at this location is potential energy. Assuming the pipe discharge point as the reference point for potential energy, the total energy for the system is:

$$E_z = \frac{zg}{g_c} = \frac{(2500\ ft - 2250\ ft)(32.2\frac{ft}{s^2})}{32.2\ \frac{lbm \times ft}{lbf \times s^2}} = 250\ ft \times lbf/lbm$$

At the pipe discharge, the potential and pressure energies are zero. Kinetic energy is the only energy present, and is used to calculate the velocity of the discharge:

$$E_T = E_v = \frac{v^2}{2g_c}; v = \sqrt{2g_c * E_T}$$

$$v = \sqrt{2\left(32.2 \frac{lbm \times ft}{lbf \times s^2}\right) \times 250 \, ft \times lbf/lbm} = 16100 \, ft/s$$

Problem

A pump is used to convey water up a gradient to an existing lake. Calculate the maximum distance the water can be pumped to if the desired discharge velocity and pressure are 7.5 ft/s and 165-lbf/in², respectively. Assume there is no fluid friction.

Solution
The pump discharge centerline will be used as the reference point for potential energy (z = 0). The total energy at this point is only due to pressure ($E_{p,1}$) and kinetic energies ($E_{v,1}$) (since the potential energy is zero where z = 0):

$$E_{T,1} = E_{p,1} + E_{v,1} = \frac{p_1}{\rho} + \frac{v_1^2}{2g_c}$$

$$E_{T,1} = \frac{165 \frac{lbf}{in^2}\left(\frac{12 \, in}{1 \, ft}\right)^2}{62.4 \, lbm/ft^3} + \frac{(7.5 \, ft/s)^2}{2\left(32.2 \frac{lbm \times ft}{lbf \times s^2}\right)}$$

$$E_{T,1} = 381.64 \, ft \times lbf/lbm$$

At the discharge point to the lake, the pressure and kinetic energy is zero, so the energy is due all to potential energy:

$$E_{T,1} = E_{T,2} = 381.64 \, ft \times lbf/lbm = E_{z,2} = \frac{z_2 g}{g_c}$$

$$z_2 = \frac{381.64 \, ft \times lbf/lbm \left(32.2 \frac{lbm \times ft}{lbf \times s^2}\right)}{32.2 \frac{ft}{s^2}} = 381.64 \, ft$$

Problem

Show how to calculate the hydraulic radius for a pipe flowing at less than half its capacity.

Solution
The hydraulic radius is the ratio of the cross-sectional flow area to the wetted perimeter, s: $r_h = A/s$. The cross-sectional flow area is calculated from:

$$A = \frac{1}{2}r^2(\phi - \sin \phi)$$

Where r is the pipe radius, and ϕ is the central angle of the pipe with respect to the flow surface, in radians. If the flow surface is a distance y below the center of the pipe, the central angle in degrees is $2 \times \arccos(y/r)$. To convert to radians, this value is multiplied by $2\pi/360°$. The wetted perimeter is the arc length, which is solved using the following equation: $s = r\phi$. The hydraulic radius is then:

$$r_h = \frac{\frac{1}{2}r^2(\phi - \sin \phi)}{r\phi}, or$$

$$r_h = \frac{\frac{1}{2}r^2\left(\left(\frac{4\pi(\arccos\ (y/r))}{360°}\right) - \sin\left(\frac{4\pi(\arccos\ (y/r))}{360°}\right)\right)}{r\left(\frac{4\pi(\arccos\ (y/r))}{360°}\right)}$$

Terms

The *Reynolds number* is a dimensionless number that is used for flow characterization. It is the ratio of the fluid's inertial forces to viscous forces.

Laminar flow is flow that is relatively smooth. This type of flow is governed by the fluid's viscous forces. The velocity distribution of laminar flow is parabolic with the velocity being zero at the point of contact with the pipe, and maximum at the center. A flow is considered laminar if the Reynolds number is less than 2100.

Turbulent flow exhibits three-dimensional movement within the pipe. This type of flow is governed by the fluid's inertial forces and is characterized by complete mixing. A flow is considered turbulent if the Reynolds number is greater than 4000.

Critical flow is the zone at which the fluid is transitioning between laminar and turbulent flows. Critical flow is unpredictable and has a Reynolds number between the bounds for laminar and turbulent flow (2100 to 4000).

Energy grade line and hydraulic grade line

The *energy grade line* is the total energy shown graphically as a function of pipe length. The elevation of the energy grade line is the sum of the pressure, velocity, and gravitational potential heads. The *hydraulic grade line* is the sum of the pressure and gravitational potential heads shown graphically with respect to pipe length. The velocity head is then the difference between the energy grade line and the hydraulic grade line at a certain point along the pipe. Without taking fluid friction into account and without the presence of pumps or turbines, the following characteristics are true: a) the energy grade line is a straight line along the pipe length, b) the hydraulic grade line would never be greater than the energy grade line, c) the hydraulic grade line decreases as the flow area decreases (and vice versa), and d) the hydraulic grade line is horizontal with conditions of constant flow velocity.

Hazen-Williams equation

The *Hazen-Williams equation* is an empirical equation used for turbulent flow primarily for pressure conduits:
$V = 1.318 C R^{0.633} S^{0.54}$ $(Units: U.S.)$
$V = 0.85 C R^{0.63} S^{0.54}$ $(Units: SI)$
Where V is the flow velocity (in feet per second or meter per second), C is the dimensionless Hazen-Williams roughness coefficient, which is a function of the material and age of the pipe, R is the hydraulic radius (in ft or m), and S is the slope of the energy gradient (in feet per foot of pipe length, or meters per meter of pipe length). For full flow conditions in a circular pipe, the Hazen-Williams equation can be restated as:
$Q = 0.279 C D^{2.63} S^{0.54}$ $(Units: U.S.)$

$Q = 0.278CD^{2.63}S^{0.54}$ (Units: SI)

Where Q is the flow (in mgd or cubic meters per second) and D is the pipe diameter (in feet or meters).

Darcy-Weisbach equation

The *Darcy-Weisbach equation* can be used for laminar or turbulent flow. The Darcy-Weisbach equation for calculating head loss is:

$$h_L = \frac{fLv^2}{2Dg}$$

Where h_L is the head loss (in feet), L is the pipe length (in feet), v is the flow velocity (in feet per second), D is the pipe diameter (in feet), f is the friction factor, and g is the gravitational constant (32.2 ft/s²). The friction factor f is related to the Reynolds number by the Colebrook equation:

$$\frac{1}{\sqrt{f}} = -2\log_{10}\left(\frac{\frac{\epsilon}{D}}{3.7} + \frac{2.5}{Re\sqrt{f}}\right)$$

Where ϵ/D is the relative roughness. A simplified equation to calculate the friction factor in ideal laminar flow is known as the Darcy equation:

$$f = \frac{64}{Re}$$

Problem

Discuss how to calculate head loss for turbulent flow conditions in which the pipe size is unknown.

Solution

The Darcy-Weisbach equation is by far the most commonly used equation to calculate head loss for turbulent flow conditions. But for situations in which the pipe size is unknown, it is difficult estimate an accurate initial value of the friction factor, and calculating head loss using this equation would likely become a long iterative process. For these situations, the Hazen-Williams equation may be used to calculate head loss as the equation does not rely on the Reynolds number:

$$h_L = \frac{3.022Lv^{1.85}}{C^{1.85}D^{1.17}}$$ (Units: U.S.)

Where C is the Hazen-Williams roughness coefficient (tabulated), L is the pipe length in feet, D is the pipe diameter in feet, and v is the flow velocity in ft/s. This equation may also be expressed as:

$$h_L = \frac{10.44Lv^{1.85}}{C^{1.85}D^{4.87}}$$ (Units: U.S.)

Where C is the Hazen-Williams roughness coefficient (tabulated), L is the pipe length in feet, D is the pipe diameter in inches, and v is the flow velocity in gpm.

Problem

Reservoir A and Reservoir B are connected by two pipes in series. The pipe exiting Reservoir A (pipe 1) has a diameter of 18 inches and a length of 2100 feet. The pipe entering Reservoir B (pipe 2) has a diameter of 12 inches and a length of 1750 feet. Calculate the velocity in pipe 2 if the total head loss is 130 feet. Neglect minor losses, and assume the friction factor f is equal to 0.01 for both pipes.

Solution
For pipes in series, the total head loss is the sum of the head loss of each section. Neglecting minor losses, the total head loss becomes:

$$h_{L,total} = \left[f_1\left(\frac{L_1}{D_1}\right)\left(\frac{v_1^2}{2g}\right)\right] + \left[f_2\left(\frac{L_2}{D_2}\right)\left(\frac{v_2^2}{2g}\right)\right]$$

$$130\ ft = \left[0.01\left(\frac{2100}{18/12}\right)\left(\frac{v_1^2}{2g}\right)\right] + \left[0.01\left(\frac{1750}{12/12}\right)\left(\frac{v_2^2}{2g}\right)\right]$$

$$\frac{v_1^2}{2g} = \frac{\left[17.5\left(\frac{v_2^2}{2g}\right)\right] - 130\ ft}{14}$$

From the continuity equation, the following relationship is true:

$$\frac{v_1^2}{2g} = \left(\frac{D_2}{D_1}\right)^4\left(\frac{v_2^2}{2g}\right) = \left(\frac{12}{18}\right)^4\left(\frac{v_2^2}{2g}\right) = 0.198\left(\frac{v_2^2}{2g}\right)$$

Substituting this value into the previous equation:

$$0.198\left(\frac{v_2^2}{2g}\right) = \frac{\left[17.5\left(\frac{v_2^2}{2g}\right)\right] - 130\ ft}{14}$$

$$v_2 = 23.84\ ft/s$$

Problem

Reservoir A and Reservoir B are connected by two pipes in series. The pipe exiting Reservoir A (pipe 1) has a diameter of 18 inches and a length of 2100 feet. The pipe entering Reservoir B (pipe 2) has a diameter of 12 inches and a length of 1750 feet. Calculate the velocity in pipe 1 and the discharge from Reservoir A to Reservoir B if the total head loss is 130 feet and the velocity in pipe 2 is 23.84 ft/s. Neglect minor losses, and assume the friction factor f is equal to 0.01 for both pipes.

Solution
From the continuity equation, the following relationship is true:

$$\frac{v_1^2}{2g} = \left(\frac{D_2}{D_1}\right)^4\left(\frac{v_2^2}{2g}\right)$$

$$v_1^2 = 2g\left(\frac{12}{18}\right)^4\left(\frac{\left(23.84\frac{ft}{s}\right)^2}{2g}\right) = 112.27\ ft^2/s^2$$

$$v_1 = 10.60\ ft/s$$

Since the flow in pipe 1 is equal to the flow in pipe 2, the conditions in either pipe may be used to calculate flow:

$$Q_1 = v_1 A_1 = 10.60\ ft/s(\pi)\left(\frac{18\ in}{2*12\ in/ft}\right)^2$$

$$Q_1 = 18.7\ ft^3/s$$

As a check, calculate Q_2 to ensure both flows are equal:

$$Q_2 = v_2 A_2 = 23.84 \, ft/s (\pi) \left(\frac{12 \, in}{2 * 12 \, in/ft}\right)^2$$

$Q_1 = 18.7 \, ft^3/s$ ✓

Problem

Three pipes are connected in parallel, as shown below. Derive an equation to calculate the total system flow based on the total head loss and the roughness coefficient, C, for each of the three pipes.

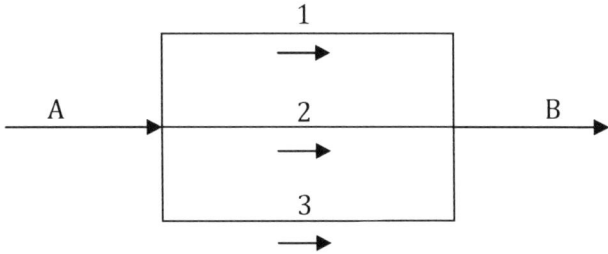

Solution
The total flow for the system is the sum of the individual flows of Pipes 1, 2 and 3. The sum is equal to both the flow entering and exiting the system:
$Q_A = Q_B = Q_1 + Q_2 + Q_3$
The head loss in each parallel pipe is equal, and is also equal to the head loss for the total system:
$h_{L,total} = h_{L,1} = h_{L,2} = h_{L,3}$
Based on the Darcy-Weisbach equation, the head loss is related to flow by:

$$Q = A \sqrt{\left(\frac{L}{D}\right) 2g h_L / f}$$

And since:

$$C = A \sqrt{\left(\frac{L}{D}\right) 2g / f}$$

The equation becomes: $Q = C(h_L)^{1/2}$

For the system shown:
$$Q = \sqrt{h_{L,total}} (C_1 + C_2 + C_3)$$

Problem

A 12-inch pipe exits a reservoir, flowing full downgrade and dumps into another reservoir. Calculate the total head loss of the system given the following:
Flow, Q = 2 cfs
Pipe Diameter, D = 8 in
Pipe Length, L = 150 ft
Friction Factor, f = 0.01
Total Minor Head Loss Coefficient, K = 5

Solution
First, calculate the velocity of the water being conveyed:
$$v = \frac{Q}{A} = \frac{2\ cfs}{\pi\left(\frac{8\ in}{2 \times 12\ in/ft}\right)^2} = 5.73\ ft/s$$
The total head loss for the system is calculated from:
$$h_{L,total} = \left[f\left(\frac{L}{D}\right)\left(\frac{v^2}{2g}\right) + \sum K\left(\frac{v^2}{2g}\right)\right]$$
Where the first term is the loss due to friction, and the second term is due to minor losses.
The head loss due to friction is:
$$h_{L,f} = 0.01\left(\frac{150}{8/12}\right)\left(\frac{(5.73\ ft/s)^2}{2(32.2\ ft/s^2)}\right) = 1.15\ ft$$
The minor head loss is:
$$h_{L,m} = \sum K\left(\frac{v^2}{2g}\right) = 5\left(\frac{(5.73\ ft/s)^2}{2(32.2\ ft/s^2)}\right) = 2.55\ ft$$
The total head loss for the system is the addition of the loss due to friction and the minor losses:
$$h_{L,total} = 1.15\ ft + 2.55\ ft = 3.7\ ft$$

Problem

A pipe network consisting of two closed loops is shown below. The branch length and diameter of each pipe are also shown. The Darcy friction factor, f, is equal to 0.02 for all network pipes. Determine the Darcy friction coefficient for each branch within the network using the Hardy Cross method.

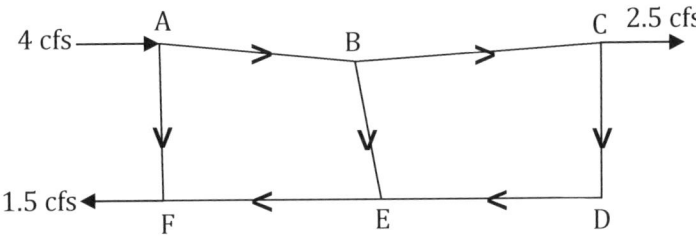

AB: L = 1000', D = 8"   **BC:** L = 2500', D = 6"
CD: L = 1550', D = 4"   **DE:** L = 2700', D = 4"
EF: L = 750', D = 4"   **BE:** L = 1700', D = 4"
AF: L = 1500', D = 4"

Solution
The Darcy friction coefficient, K' for each branch can be calculated from the following equation:
$$K' = \frac{0.02517 \times f \times L}{D^5}$$
Where f is the Darcy friction factor, L is the pipe branch length in feet and D is the diameter of the pipe branch in feet.

$$K'_{AB} = \frac{0.02517(0.02)(1000\ ft)}{\left(\frac{8}{12} ft\right)^5} = 3.8$$

$$K'_{BC} = \frac{0.02517(0.02)(2500\ ft)}{\left(\frac{6}{12} ft\right)^5} = 40.3$$

Solving similarly for the remaining branches results in the following:
$K'_{CD} = 189.6, \quad K'_{DE} = 330.3$
$K'_{EF} = 91.7, \quad K'_{BE} = 208.0$
$K'_{AF} = 183.5$

Problem

A pipe network consisting of two closed loops is shown below. The external flows into/out of nodes A, C, and F are known. The flows within the loops have been assumed, including direction of flow. Given the Darcy friction coefficients for each branch within the network (f = 0.02), calculate the flow correction for the loop ABEF and loop BCDE using the Hardy Cross method.

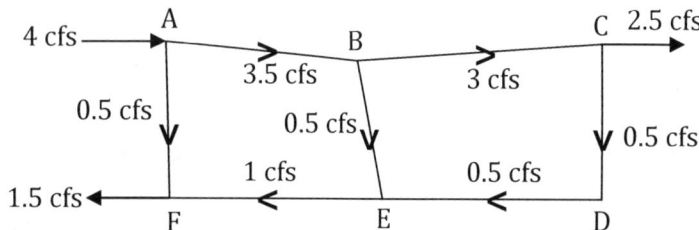

AB: $K'_{AB} = 3.8$ **BC**: $K'_{BC} = 40.3$
CD: $K'_{CD} = 189.6$ **DE**: $K'_{DE} = 330.3$
EF: $K'_{EF} = 91.7$ **BE**: $K'_{BE} = 208.0$
AF: $K'_{AF} = 183.5$

Solution
Assume flow in clockwise direction is positive. The flow correction for a loop, δ, can be calculated from:
$$\delta = \frac{-\sum K' v_a^n}{n \sum |K' v_a^{n-1}|}$$
Where K' is the Darcy friction coefficient, v_a is the assumed branch velocity in cfs, and n an exponent (n = 2 for a Darcy head loss). The correction for loop ABEF is:
$\delta_{ABEF} =$
$$\frac{-[(3.8)(3.5)^2 + (208)(0.5)^2 + (91.7)(1)^2 - (183.5)(0.5)^2]}{2[(3.8)(3.5) + (208)(0.5) + (91.7)(1) + (183.5)(0.5)]}$$
$\delta_{ABEF} = -0.24$
And the correction for loop BCDE is:
$\delta_{BCDE} =$
$$\frac{-[(40.3)(3)^2 + (189.6)(0.5)^2 + (330.3)(0.5)^2 - (208)(0.5)^2]}{2[(40.3)(3) + (189.6)(0.5) + (330.3)(0.5) + (208)(0.5)]}$$
$\delta_{BCDE} = -0.45$

Problem

Calculate the theoretical and actual orifice jet velocity for a non-pressurized tank with a midpoint of discharge 13.5 feet below the tank water surface. Assume the orifice is sharp-edged.

Solution
The theoretical orifice jet velocity, v_t, can be calculated from *Torricelli's speed of efflux* equation:
$$v_t = \sqrt{2gh}$$
Where h is the difference in elevation between the tank water surface and the midpoint of the discharge exit. The theoretical velocity is then:

$$v_t = \sqrt{2\left(32.2 \frac{ft}{s^2}\right) 13.5 \, ft} = 29.5 \, ft/s$$

The actual orifice jet velocity is calculated from:
$$v_{actual} = C_v\sqrt{2gh}$$
Where C_v is the velocity coefficient. For a sharp-edged orifice, the velocity coefficient is 0.98.

$$v_{actual} = 0.98 * \sqrt{2\left(32.2 \frac{ft}{s^2}\right) 13.5 \, ft} = 28.9 \, ft/s$$

Problem

Calculate the time to empty a tank of constant area of 314.16 ft² having an initial water surface elevation 15 feet above a round-edged (C_d = 0.98), bottom orifice with an area of 0.0218 ft². Assume the tank is subjected to a constant pressure of 65 lbf/in².

Solution
The initial head, h_1, that causes the discharge is calculated from the following equation:
$$h_1 = z_1 + \left(\frac{p}{\rho}\right)\left(\frac{g_c}{g}\right) = z_1 + h_2$$

$$h_1 = 15 \, ft + \left(\frac{(65 \, lbf/in^2) 144 \, in^2/ft^2}{61.2 \, lbm/ft^3}\right)\left(\frac{32.2 \frac{ft \times lbm}{lbf \times s^2}}{32.2 \, ft/s^2}\right)$$

$h_1 = 167.9 \, ft$
At the orifice level, the final head, h_2 is then = 152.9 ft.
The time to empty the tank is then calculated from:
$$t = \frac{2A_{tank}\left(\sqrt{h_1} - \sqrt{h_2}\right)}{C_d A_{orifice}\sqrt{2g}}$$

$$t = \frac{2(314.16 \, ft)\left(\sqrt{167.9 \, ft^2} - \sqrt{152.9 \, ft}\right)}{(0.98)(0.0128 \, ft^2)\sqrt{2(32.2 \, ft/s^2)}} = 3697 \, s = 61.6 \, min$$

$t = 3697 \, s = 61.6 \, min$

Problem

Water at a rate of 50 gpm is pumped uphill a total distance of 375 feet. Calculate the motor size required, assuming the pump efficiency is 66%. Also calculate the cost required to operate this pump for 12 hours, assuming a motor efficiency of 87% and an electricity cost of $0.10/kW-hr.

Solution
The motor power requirement, P (in horsepower), can be calculated from:
$$P = \frac{Q h_p SG}{3956 \eta_p}$$
Where Q is the flow rate in gpm, h_p is the head developed by the pump in feet, SG is the specific gravity of the liquid being pumped, and η_p is the pump efficiency.
$$P = \frac{(50 \ gpm)(375 \ ft)(1)}{3956(0.66)} = 7.18 \ hp$$
The next highest standard motor size is 7.5 hp, which would be the size motor selected.
The cost to operate the pump for 12 hours is calculated from:
$$Cost = (Cost \ per \ kW * hr) \frac{Pt}{\eta_m}$$
Where t is the time of operation and η_m is the motor efficiency.
$$Cost = \left(\frac{\$0.10}{kW \times hr}\right) \frac{7.18 \ hp (12 \ hours)}{0.87} = \$9.90$$

Positive displacement pumps and kinetic pumps

Positive displacement pumps cause a fixed volume of flow to be moved (displaced) into the discharge pipe with each stroke or revolution of the pump. This method of conveyance results in intermittent energy being added to the fluid system. Two types of positive displacement pumps are rotary action pumps and reciprocating actions pumps.

Kinetic pumps are pumps that convert kinetic energy to static pressure energy in the fluid. This method results in continuous energy within the fluid system. A common type of kinetic pump is the centrifugal pump. With centrifugal pumps, a rotating impeller moves the suction water against the casing, creating an increase in velocity which is converted to static pressure energy, and discharges the water into the exit pipe. A common characteristic of centrifugal pumps is that the suction pipe is usually at a right angle to the discharge pipe. Centrifugal pumps are typically used for pumping sewage.

Problem

The base of an elevated open-air water tank is located 20 feet above grade. The tank is filled up with 8 feet of water at 70 °F and standard atmospheric pressure (14.7 psi). The water from the tank is pumped to another open-air tank via a pump with an inlet elevation located 3 feet above grade. Calculate the net positive suction head available if the total loss due to friction in the suction piping and fittings is 3.1 feet.

Solution
Given the information on the suction side of an open reservoir fluid surface, the following equation can be used to calculate the actual energy of the fluid at the pump suction side:
$NPSHA = h_{atm} + h_{z,suction} - h_{f,suction} - h_{vp}$
Where h_{atm} is the atmospheric head, $h_{z,suction}$ is the static head, $h_{f,suction}$ is the head loss due to friction, and h_{vp} is the vapor pressure head loss (all terms in reference to the suction side). The term h_{vp} is tabulated; at 70°F, the vapor pressure head is 0.84 feet. The remaining terms are calculated from:

$h_{atm} = \dfrac{p}{\gamma} = \dfrac{\left(14.7 \frac{lbf}{in^2}\right)\left(144 \frac{in^2}{ft^2}\right)}{62.4 \, lbf/ft^3} = 33.9 \, ft$

$h_{z,suction} = 8 \, ft + 20 \, ft - 3 \, ft = 25 \, ft$

$h_{f,suction} = 3.1 \, ft \, (given)$

The net positive suction head available is then:
$$NPSHA = 33.9 \, ft + 25 \, ft - 3.1 \, ft - 0.84 \, ft = 55.0 \, ft$$

Problem

Describe a pump system curve for a single pump as compared to parallel pumps.

Solution
A system curve is a plot of the total dynamic head (TDH) of the system (the sum of the static and friction energy losses in the system) plotted along the y-axis at varying flow rates (flow along the x-axis). The system curve is plotted against manufacturer pump curves. The point of intersection between the two curves is called the operating point. It identifies the system head and flow rate that the pump is capable of for the system curve given. For pumps in parallel, the combined pump curve is plotted by adding the flow rate capacities of each pump at a given head (then doing the same at various heads). The system curve is plotted against the individual pump curves and the combined pump curve. The point at which the system curve intersects the combined pump curve is the operating point for the condition at which both/all of the parallel pumps are pumping.

Manning equation

The *Manning equation* (also known as the Chezy-Manning equation) is shown below:

$v = \dfrac{1.00}{n} R^{\frac{2}{3}} \sqrt{S}$ (Units: SI)

$v = \dfrac{1.49}{n} R^{\frac{2}{3}} \sqrt{S}$ (Units: U.S.)

Where v is the flow velocity in meters per second or feet per second, n is Manning's roughness coefficient (tabulated), R is the hydraulic radius in meters or feet, and S is the slope of the energy gradient (which will equal the geometric slope of the channel for conditions of uniform flow). When the equation $Q = vA$ is substituted into the Manning equation and solved for Q, the following equation results:

$Q = \dfrac{1.00}{n} A R^{\frac{2}{3}} \sqrt{S}$ (Units: SI)

$Q = \dfrac{1.49}{n} A R^{\frac{2}{3}} \sqrt{S}$ (Units: U.S.)

Which can then be reduced to $Q = K\sqrt{S}$, where K is the conveyance.

Problem

Define the variables in the Chezy equation used to calculate the velocity of open channel flow.

Solution
The velocity of open channel flow is most commonly calculated from the following Chezy equation:
$$v = C\sqrt{RS}$$
Where v is the flow velocity, C is the Chezy coefficient, R is the hydraulic radius, and S is the slope of the energy gradient (which will equal the geometric slope of the channel for conditions of uniform flow). For small and smooth channels, the Chezy coefficient may be calculated from:
$$C = \sqrt{\frac{8g}{f}}$$
Where f is the Reynold's number-dependent friction factor. For large channels with fully turbulent flow, the Chezy coefficient may be calculated from:
$$C = \left(\frac{1.00}{n}\right) R^{1/6} \quad (Units: SI)$$
$$C = \left(\frac{1.49}{n}\right) R^{1/6} \quad (Units: U.S.)$$
Where n is Manning's roughness coefficient (tabulated).

Problem

Calculate the discharge of a trapezoidal channel having a grass bottom and sides (n = 0.027) for the given information below:
Normal depth of flow, d_n = 8 feet
Open channel bottom width = 14 feet
Open channel water surface width = 22 feet
Channel slope = 0.002

Solution
Since the normal depth is given, it is known that the flow is uniform and the slope of the energy gradient S is also equal to the channel slope. The hydraulic radius of a trapezoid is shown below:
$$R = \frac{bd \sin\theta + d^2 \cos\theta}{b \sin\theta + 2d}$$
The outer angle between the channel bottom and earth side is equal to:
$$\tan\theta = \frac{8}{(22-14)/2}; \quad \theta = 63.43°$$
$$R = \frac{(14)(8)\sin 63.43° + (8)^2 \cos 63.43°}{(14)\sin 63.43° + 2(8)} = 4.5 \, ft$$
The area of trapezoid is:
$$A = d\left(b + \frac{d}{\tan\theta}\right) = 8\left(14 + \frac{8}{\tan 63.43°}\right) = 144.0 \, ft^2$$
$$Q = \frac{1.49}{n} A R^{\frac{2}{3}} \sqrt{S} = \frac{1.49}{0.027}(144 \, ft^2)(4.5 \, ft)^{\frac{2}{3}}\sqrt{0.002}$$
$$Q = 968.7 \, cfs$$

Problem

A 12-inch gravity sewer pipe is flowing partially full at a flow of 0.83 cfs, with a Manning's roughness coefficient of 0.014 and at a slope of 0.002. Calculate the depth of flow assuming Manning's roughness coefficient varies with depth.

Solution
First calculate the hydraulic radius and flow for a 12-inch pipe flowing full. For full flow in a circular pipe, the hydraulic radius is:

$$R = \frac{D}{4} = \frac{12 \text{ in } (1 \text{ ft}/12 \text{ in})}{4} = 0.25 \text{ ft}$$

The flow with the pipe flowing full can be calculated using the full flow hydraulic radius and the full pipe area ($\pi r^2 = \pi(0.5 \text{ ft})^2 = 0.785 \text{ ft}^2$):

$$Q_{full} = \frac{1.49}{n} AR^{\frac{2}{3}}\sqrt{S} = \frac{1.49}{0.014}(0.785 \text{ ft}^2)(0.25 \text{ ft})^{\frac{2}{3}}\sqrt{0.002}$$

$$Q_{full} = 1.48 \text{ cfs}; \quad \frac{Q}{Q_{full}} = \frac{0.83 \text{ cfs}}{1.48 \text{ cfs}} = 0.56$$

For cases of Manning's roughness coefficient, n, varying with depth, use tabulated or graphical data to determine d/D. For a Q/Q_{full} value of 0.56, the d/D value is 0.6. The depth of flow is then:

$$d = D * 0.6 = 12 \text{ in } \left(\frac{1 \text{ ft}}{12 \text{ in}}\right)(0.6) = 0.6 \text{ ft}$$

Problem

The depths of flow upstream and downstream a 15-ft wide sluice gate are 10 feet and 3 feet, respectively. Calculate the upstream velocity, assuming the flow is uniform and the channel bottom is level.

Solution
Since the bottom of the channel is level, the terms z_1, z_2, and h_f can all be eliminated from the Bernoulli equation, leaving:

$$\frac{v_1^2}{2g} + \frac{p_1}{\rho} = \frac{v_2^2}{2g} + \frac{p_2}{\rho}$$

And since $p_1/\rho = d_1$ and $p_2/\rho = d_2$, the equation can be rewritten as:

$$\frac{v_1^2}{2g} + d_1 = \frac{v_2^2}{2g} + d_2 \Rightarrow \frac{v_1^2}{2g} + 10 \text{ ft} = \frac{v_2^2}{2g} + 3 \text{ ft}$$

And from the continuity equation, $v_1 A_1 = v_2 A_2$. Solving for v_2: $v_2 = (v_1 A_1)/A_2$. A_1 is equal to (15 ft)(10 ft) = 150 sf. A_2 is equal to (15 ft)(3 ft) = 45 sf. So $v_2 = 3.33(v_1)$. Substituting this into the previous equation:

$$\frac{v_1^2}{2g} + 10 \text{ ft} = \frac{(3.33v_1)^2}{2g} + 3 \text{ ft}$$

$$0.157v_1^2 = 7 \text{ ft}$$

$$v_1 = 6.7 \text{ ft/s}$$

Problem

Calculate the energy loss per 1000 feet of normal flow in a channel given the following:
Manning's roughness coefficient, n = 0.013
Energy gradient slope, S = 0.001
Hydraulic radius, R = 4.8 feet
Flow velocity, v = 7.15 feet per second

Solution
The energy loss can be calculated from the following equation:
$$h_f = \frac{Ln^2 v^2}{2.208 R^{4/3}}$$
Where L is the channel length, n is Manning's roughness coefficient, v is the flow velocity, and R is the hydraulic radius. The energy loss per 1000 feet is then:
$$h_f = \frac{(1000\ ft)(0.013)^2(10.285\ ft/s)^2}{2.208(4.80\ ft)^{4/3}} = 1\ ft$$
As a check, the energy loss can also be calculated from the following equation:
$$h_f = LS$$
Where S is the energy gradient slope (equal to the channel slope for uniform flow). From this equation the energy loss per 100 feet is:
$$h_f = (1000\ ft)(0.001) = 1\ ft\ \checkmark$$

Problem

Design the most efficient dimensions for a brick (n = 0.015) rectangular open channel to carry 850 cfs at a channel slope of 0.002. Assume uniform flow conditions.

Solution
The most efficient dimensions for a rectangle are such that the channel width, w, is twice the channel depth, d: $w = 2d$. Solving for the channel area: $A = 2d \times d = 2d^2$. The wetted perimeter, P, is then equal to $2d + d + d = 4d$. The hydraulic radius, R, is calculated from:
$$R = \frac{A}{P} = \frac{2d^2}{4d} = 0.5d$$
Plugging in the area, hydraulic radius, and slope into the Manning equation, and solving for d:
$$Q = \frac{1.49}{n} A R^{\frac{2}{3}} \sqrt{S}$$
$$850\ cfs = \frac{1.49}{0.015}(2\ d^2)(0.5\ d)^{\frac{2}{3}}\sqrt{0.002}$$
$$850\ cfs = 5.60\ cfs \times d^{8/3}$$
$$d = 6.58\ ft$$
And solving for the channel width:
$$w = 2d = 2 \times 6.58\ ft = 13.16\ ft$$

Culvert flow

Culverts are defined as pipes (or barrels) that bypass obstructions by transporting water through or under structures. Culverts are either inlet- or outlet-controlled. The flow into an inlet-controlled culvert is characterized by being slower than the flow going through and out of the culvert. These types of culverts are never full, only partially full. The flow into an outlet-controlled culvert is characterized by being faster than the flow going through and out of the culvert. Outlet-controlled culverts may either flow full or partially full. Culvert entrances and exits may be partially or fully submerged. Culvert exits may also consist of a free outfall. There are six different types of culvert flow classifications. These classifications are based on parameters such as control type and location, pipe slope, flow type, and headwater and tailwater heights.

Culvert flow classification type 1

Culvert flow type 1 characteristics include the following: a steep pipe slope, partially full flow, inlet-controlled, water passes through critical depth near culvert inlet, and the tailwater height is less than the critical height. For this type of culvert, the flow can be calculated from:

$$Q_{T1} = C_d A_c \sqrt{2g\left(h_1 - z + \frac{\alpha v_1^2}{2g} - d_c - h_{f,1-2}\right)}$$

Where C_d is a discharge coefficient, A_c is the area of flow at critical depth of flow, h_1 is the depth of water at inlet (difference between water surface elevation and elevation of outlet), z is the difference between the inlet and outlet elevations, α is a velocity-head coefficient (Coriolis coefficient), v_1 is the average approach velocity, d_c is the actual depth of flow within the culvert, and $h_{f,1-2}$ is the friction loss across the culvert inlet.

Culvert flow classification type 2

Culvert flow type 2 characteristics include the following: a mild pipe slope, partially full flow, outlet-controlled, water passes through critical depth near culvert outlet, and the tailwater height is less than the critical height. For this type of culvert, the flow can be calculated from:

$$Q_{T2} = C_d A_c \sqrt{2g\left(h_1 + \frac{\alpha v_1^2}{2g} - d_c - h_{f,1-2} - h_{f,2-3}\right)}$$

Where C_d is a discharge coefficient, A_c is the area of flow at critical depth of flow, h_1 is the depth of water at inlet (difference between water surface elevation and elevation of outlet), α is a velocity-head coefficient (Coriolis coefficient), v_1 is the average approach velocity, d_c is the actual depth of flow within the culvert, $h_{f,1-2}$ is the friction loss across the culvert inlet, and $h_{f,2-3}$ is the friction loss across the length of the culvert.

Problem

Calculate the discharge for the two scenarios below:
a) Broad-crested weir:
Water height, H, over weir = 5 inches
Base width of weir, b = 10 feet
Flow coefficient, C_1 = 0.56

b) Ogee spillway:
Water height, H, over spillway = 5 inches
Base width of spillway, b = 10 feet

Spillway coefficient, $C_s = 3.97$ ft$^{0.5}$/s
Assume approach velocity is insignificant.

Solution
The following equation can be used to calculate the discharge for broad-crested weirs and for ogee spillways:
$$Q = \frac{2}{3}bC_1 H^{3/2}\sqrt{2g}$$
$$Q = \frac{2}{3}(10\ ft)(0.56)\left(5\ in\left(\frac{1\ ft}{12\ in}\right)\right)^{3/2}\sqrt{2\left(32.2\frac{ft}{s^2}\right)}$$
$$Q = 8.06\ cfs$$
The following equation can be used for either broad-crested weirs or for ogee spillways as long as the approach velocity is considered insignificant:
$$Q = bC_s H^{3/2}$$
$$Q = (10\ ft)(3.97 ft^{0.5}/s)\left(5\ in\left(\frac{1\ ft}{12\ in}\right)\right)^{3/2}$$
$$Q = 10.68\ cfs$$

Hydraulic jump vs. hydraulic drop

A *hydraulic jump* occurs when water at a high velocity and low depth (supercritical flow) is introduced to water with a low velocity and high depth (subcritical flow). At the point where the two meet, the velocity will be rapidly reduced and the water level increased, causing an abrupt rise in the water surface. This is known as a hydraulic jump.

A *hydraulic drop* occurs when the opposite conditions arise. When water with a low velocity and high depth (subcritical flow) meets water with a high velocity and low depth (supercritical flow), the velocity greatly increases and an abrupt drop in the water surface is evidenced. This is known as a hydraulic drop.

Problem

Calculate the depth of a hydraulic jump if the velocity and depth within an open rectangular channel prior to the jump are 6.24 m/s and 0.0734 meters per second, respectively.

Solution
The equation that relates the depths before and after a hydraulic jump is shown below:
$$\frac{y_b}{y_a} = \frac{\sqrt{1+8Fr_a^2}-1}{2}$$

Where y_a is the water depth prior to the hydraulic jump, y_b is the water depth after the hydraulic jump, and Fr_a is the Froude number upstream of the hydraulic jump. The Froude number can be calculated from:

$$Fr_a = \frac{v_a}{\sqrt{gy_a}}$$

Where v_a is the velocity upstream of the jump.

$$Fr_a = \frac{6.24 \ m/s}{\sqrt{9.81 m/s^2 (0.0734 \ m)}} = 7.35$$

$$\frac{y_b}{0.0734 \ m} = \frac{\sqrt{1 + 8(7.35)^2} - 1}{2} \ ; \ y_b = 0.727 \ m$$

Problem

The invert elevation of Pipe 1 leaving Manhole A is 253.72 ft. Calculate the invert elevations of Pipes 1 and 2 at Manholes B and C. Assume a 0.02 ft drop across each manhole.

Pipe	Pipe Length	Pipe Slope	From Manhole	To Manhole
1	375 ft	0.006	A	B
2	280 ft	0.004	B	C

Solution
Starting with the invert elevation of Pipe 1 at Manhole A given as 253.72 ft, determine the elevation drop due to slope to point of connection to Manhole B:
$Elevation \ Drop = Pipe \ Length \times Pipe \ Slope$
$Elevation \ Drop_{Pipe \ 1} = 375 \ ft \times 0.006 = 2.25 \ ft$
The invert elevation of Pipe 1 at Manhole B is then:
$IE_{Pipe \ 1 \ @ \ MHB} = 253.72 \ ft - 2.25 \ ft = 251.47 \ ft$
The invert of Pipe 2 leaving Manhole B is calculated by subtracting the elevation drop across Manhole B: 251.47 ft – 0.02 ft = 251.45 ft.
The elevation drop due to slope for Pipe 2 at point of connection to Manhole C is:
$Elevation \ Drop_{Pipe \ 2} = 280 \ ft \times 0.004 = 1.12 \ ft$
The invert elevation of Pipe 2 at Manhole C is then:
$$IE_{Pipe \ 2 \ @ \ MHC} = 251.45 \ ft - 1.12 \ ft = 250.33 \ ft$$

Problem

A stormwater detention pond is used during storm events to capture and divert the stormwater for future irrigation purposes. If the outflow exits the detention pond via a pipe, derive an equation for detention pond depth as a function of time.

Solution
The change in the detention pond volume, or change in storage, S, over time is based on the inflow and outflow to and from the pond:

$$Q_{in} - Q_{out} = \frac{dS}{dt}$$

The change in storage over time can also be expressed as the product of the mean surface area, A, at a given depth, h, and the depth increment:

$$\frac{dS}{dt} = Ah\frac{dh}{dt}$$

Since the outflow from the pond exits via a pipe, Q_{out} can be expressed as:

$$Q_{out} = C_d A_o (2gh)^{0.5}$$

Where C_d is a discharge coefficient and A_o is the area of the orifice/pipe. Plugging the two latter equations into the first equation and solving for dh/dt, the following is obtained:

$$\frac{dh}{dt} = \frac{Q_{in} - C_d A_o (2gh)^{0.5}}{Ah}$$

Terms

A *100-year flood* is a flood that has a 1% (or greater) chance of occurring within a year. A 100-year flood is also known as a base flood.

A *500-year flood* is a flood that has a 0.2% (or greater) probability of occurring within a year.

Base flood elevation, BFE, is the elevation of the 100-year flood water surface with respect to a specified datum (i.e., mean sea level).

The *design flood* is the flood that is selected (typically based on economics) specifically for the conditions of a site and used for the site's project design.

Probable maximum precipitation, PMP, (also known as probable maximum flood) is the greatest possible depth of precipitation estimated to occur over a specified duration and season for a particular geographical area. Generally, the recurrence interval of the PMP is substantially greater than 100 years.

Problem

Calculate the probability of the following:
A 100-year flood occurring in 60 years
A 500-year flood occurring in 60 years

Solution
The equation to calculate the probability of an event in a certain number of years is shown below:

$$p\{F \text{ event in } n \text{ years}\} = 1 - \left(1 - \frac{1}{F}\right)^n$$

Where F is the event and n is the number of years. The probability of a 100-year flood (same as 1% flood) occurring within 60 years is then:

$$p\{100 \text{ year flood in } 60 \text{ years}\} = 1 - \left(1 - \frac{1}{100}\right)^{60}$$

$$p\{100 \text{ year flood in } 60 \text{ years}\} = 0.45 = 45\%$$

The probability of a 500-year flood (same as 0.2% flood) occurring within 60 years is:

$$p\{500 \text{ year flood in } 60 \text{ years}\} = 1 - \left(1 - \frac{1}{500}\right)^{60}$$

$$p\{500 \text{ year flood in } 60 \text{ years}\} = 0.11 = 11\%$$

Sharp-crested weirs

A *sharp-crested weir* is an intended structure placed in a channel, so that the water flows over the weir or through the weir opening, and the flow may then be measured. Weir openings are typically rectangular, although V-notch (triangular) and trapezoidal weirs are also used. Weir openings may extend the entire width of the channel or may only take up a portion of the channel width. For the latter case, the weir is considered contracted, since the flow, or nappe, contracts as it falls over the weir (for non-submerged weir). Weirs are considered suppressed if the weir opening extends the entire channel width (as in the nappe contractions are suppressed). Weirs may also be submerged, such that the upstream height above the weir will be higher than the downstream height above the weir. Calculating flows with submerged weirs in the field is often difficult due to the need to accurately determine the upstream and downstream heights.

Problem

Calculate the flow through the weir for the two scenarios below:
a) V-notch 90° weir:
Water height, H, through weir = 5 inches
Flow coefficient, C_t = 0.593

b) Trapezoidal weir:
Water height, H, through weir = 5 inches
Base width of weir, b = 9 inches

Solution
To calculate the flow through a V-notch weir, the following equation can be used:

$$Q = \frac{8}{15} C_t \tan\left(\frac{\theta}{2}\right) H^{5/2} \sqrt{2g}$$

$$Q = \frac{8}{15}(0.593)\tan\left(\frac{90°}{2}\right)\left(5\ in\left(\frac{1\ ft}{12\ in}\right)\right)^{5/2} \sqrt{2\left(32.2\frac{ft}{s^2}\right)}$$

$Q = 0.28\ cfs$

b) To calculate the flow through a trapezoidal weir, the following equation can be used:

$$Q = 3.367 b H^{\frac{3}{2}}$$

$$Q = 3.367(9\ in)\left(\frac{1\ ft}{12\ in}\right)\left(5\ in\left(\frac{1\ ft}{12\ in}\right)\right)^{3/2}$$

$Q = 0.68\ cfs$

Problem

Calculate the difference in flow between two Parshall flumes, one with a throat width of 0.50 feet and the other flume with a 1.0 foot throat width. The design height of the water upstream of the flume throat, H_a, is 4.5 feet in both cases.

Solution
The equation to calculate the flow in a Parshall flume is shown below:

$$Q = KbH_a^n$$

Where K is a coefficient based on the Parshall flume throat width and b is the flume throat width. The value of n can be calculated from:
$$n = 1.522b^{0.026}$$

First, calculate the n values for both scenarios:
$$n_{0.5} = 1.522(0.5\ ft)^{0.026} = 1.49$$
$$n_{1.0} = 1.522(1.0\ ft)^{0.026} = 1.52$$

The values of the coefficient K are tabulated: the K values for a 0.5 ft throat width and a 1.0 ft throat width are 4.12 and 4.00, respectively. The flows can then be calculated:
$$Q_{0.5} = (4.12)(0.5\ ft)(4.5\ ft)^{1.49} = 19.37\ cfs$$
$$Q_{1.0} = (4.00)(1.0\ ft)(4.5\ ft)^{1.52} = 39.35\ cfs$$

The difference in flow between the two flumes is then:
$$Q_{1.0} - Q_{0.5} = 39.35\ cfs - 19.37\ cfs = 19.98\ cfs$$

Problem

A 250-acre lake has the following data collected over a one-month (30-day) period:
Outflow from lake, Q_{out} = 12.1 cfs
Recorded precipitation, P = 2.2 inches
Change in storage, ΔS = increase of 13 acre-feet
Evaporation, E = 11.4 inches

Calculate the inflow to the lake for that month, assuming infiltration is insignificant.

Solution
The simple water balance equation for a surface water system is shown below:
$$Q_{in} - Q_{out} + P - E - I = \Delta S$$

Where Q_{in} and Q_{out} are the inflow and outflow, respectively, P is the precipitation, E is evaporation, I is infiltration, and ΔS is the change in storage. Solving for Q_{in} (and neglecting infiltration), the equation becomes:
$$Q_{in} = \Delta S + Q_{out} - P + E$$

The outflow and change in storage must be converted to units of depth:
$$Q_{out} = \frac{12.1 \frac{ft^3}{s} \left(\frac{1\ acre}{43{,}560\ ft^2}\right)\left(\frac{12\ in}{1\ ft}\right)(1\ month)\left(\frac{2{,}592{,}000\ s}{1\ month}\right)}{250\ acre}$$
$$Q_{out} = 35.56\ inches$$
$$\Delta S = \frac{(13\ acre \times ft)(12\ in/1\ ft)}{250\ acre} = 0.62\ in$$

The inflow in inches is then:
$$Q_{in} = 0.62\ in + 35.56\ in - 2.2\ in + 11.4 = 45.38\ in$$

Problem

A 9,500 acre watershed is drained by a river carrying an average annual flow of 20 cfs. Over a one-year period, the precipitation for the watershed was recorded as 54.8 inches. Assume that the water levels in the watershed are the same at the beginning and end of the one-year period. Calculate the annual combined effects due to groundwater infiltration, evaporation, and transpiration. Also, calculate the runoff coefficient.

Solution
The water balance equation for a watershed is:
$\Delta S = P - R - G - E - T$
Where ΔS is the change in storage, P is the precipitation, R is runoff, G is groundwater infiltration, E is evaporation, and T is transpiration. Because the watershed levels are the same at the beginning and end of the year, the change in storage is 0. Solving for $G + E + T$, the equation becomes:
$G + E + T = P - R$
The runoff in units of inches is:

$$R = \frac{20 \frac{ft^3}{s} \left(\frac{1\ acre}{43{,}560\ ft^2}\right)\left(\frac{12\ in}{1\ ft}\right)(1\ year)\left(\frac{3.15 \times 10^7\ s}{1\ year}\right)}{9{,}500\ acre}$$

$R = 18.27\ inches$
$G + E + T = 54.8\ in - 18.27\ in = 36.5\ inches$
The runoff coefficient is the runoff divided by precipitation:

$$Runoff\ Coefficient = \frac{R}{P} = \frac{18.27\ in}{54.8\ in} = 0.33$$

Problem

The data shown below was collected during a storm event. For the given information, sketch the hyetograph for the storm event.

Time Period	Measured Precipitation (in)
7:00 – 7:30a	0.25 in
7:30 – 8:00a	0.75 in
8:00 – 8:30a	1.00 in
8:30 – 9:00a	1.00 in
9:00 – 9:30a	1.50 in
9:30 – 10:00a	1.75 in
10:00 – 10:30a	1.50 in
10:30 – 11:00a	1.00 in
11:00 – 11:30a	0.50 in
11:30 – 12:00p	0.25 in
12:00 – 12:30p	0.25 in
12:30 – 1:00p	0 in

Solution
A hyetograph is a graph that indicates the instantaneous rainfall intensity as evidenced within a given time period of the storm event. The hyetograph for the storm event is shown below:

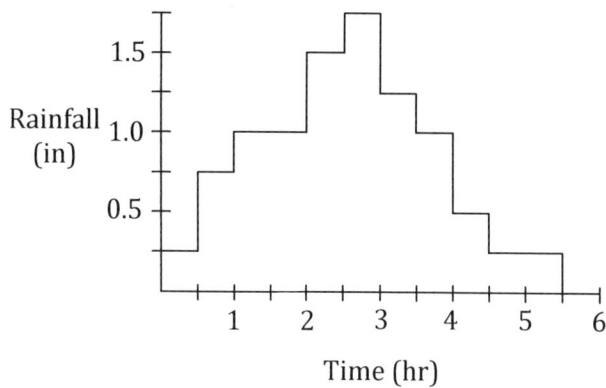

Problem

Define a hydrograph and label the typical hydrograph components on the following graph:

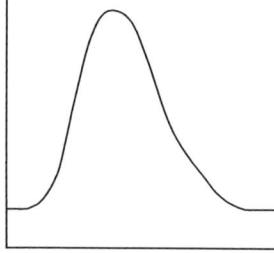

Solution

A hydrograph is a plot of instantaneous stream discharge as a function of time. The base flow is the original flow that is evidenced prior to and after the storm event. Base flow is commonly subtracted from the total hydrograph to obtain a direct runoff curve. The direct runoff (area under the curve minus base flow) is equal to the net rainfall volume.

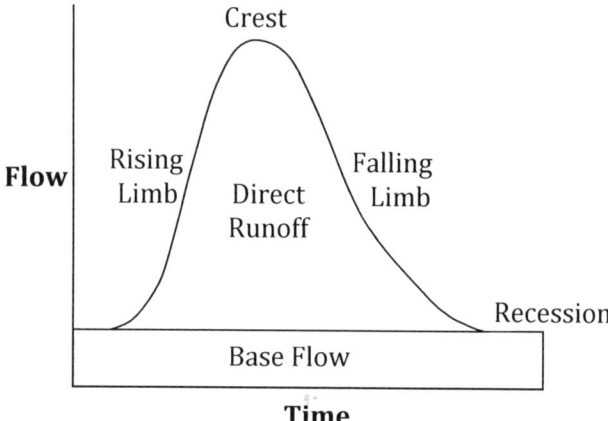

Problem

Flow in cubic feet per second is shown against time in hours in the direct runoff hydrograph and the corresponding bar graph. Determine the total volume of runoff over the watershed.

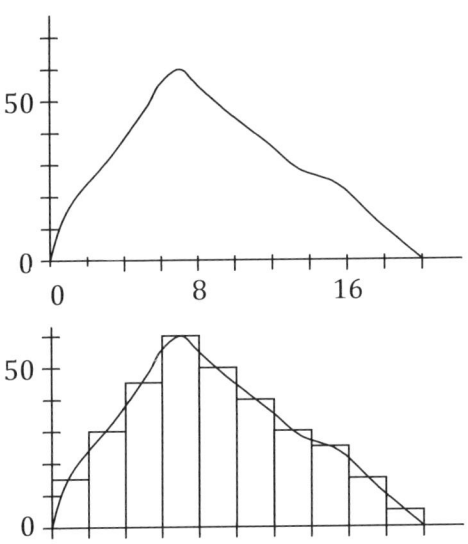

Solution
The direct runoff volume can be computed by summing the individual bar areas. The average flow is multiplied by the time to obtain the volume within that bar. The total volume is then 630 cfs*hr or 630 acre*in.

Time (hours)	Ave Flow (cfs)	Volume (cfs*hr)
0-2	15	30
2-4	30	60
4-6	45	90
6-8	60	120
8-10	50	100
10-12	40	80
12-14	30	60
14-16	25	50
16-18	15	30
18-20	5	10
	Total Volume	630

Problem

A watershed is located in central Oregon. The time it takes for the runoff from the furthest point in the watershed to reach the watershed discharge pipe is 35 minutes. Use the Steel formula to calculate the intensities of a 25-year frequency storm and a 100-year frequency storm for this watershed.

Solution
The Steel equation to calculate the intensity of a storm, I, is given below:
$$I = \frac{K}{t_c + b}$$
Where t_c is the time of concentration and K and b are tabulated coefficients dependent on the watershed's location and the frequency of the storm. Oregon is located in Region 7, and

for a 25-year storm, the coefficients K and b are 67 in*min/hr and 10 min, respectively. For a 50-year storm in Region 7, coefficients K and b are 77 in*min/hr and 10 min, respectively. The time of concentration is equal to the time it takes for the furthest point in the watershed to reach the watershed drainage point. Thus, the intensities for the 25-year and 100-year storm event are:

$$I_{25yr} = \frac{67 \text{ in} \times \text{min/hr}}{35 \text{ min} + 10 \text{ min}} = 1.5 \text{ in/hr}$$

$$I_{100yr} = \frac{77 \text{ in} \times \text{min/hr}}{35 \text{ min} + 10 \text{ min}} = 1.7 \text{ in/hr}$$

Problem

A hydrograph is developed for a 0.30 square mile (8.36 x 10^6 ft^2) watershed from the direct runoff data shown below. Determine the new flow ordinate values that would be used to develop the unit hydrograph for this storm event.

Time (hours)	Q (cfs)
0-2	30
2-4	45
4-6	42
6-8	40
8-10	33
10-12	26

Solution

A unit hydrograph is developed by reducing the hydrograph flow coordinates by a factor equal to the average depth of precipitation over the watershed, which is equal to the volume of total watershed runoff divided by the watershed area. The runoff volume is calculated as cubic feet and totaled as shown in the table. The average depth of precipitation is then = (1,555,200 ft^3)(12 in/1 ft)/(8.36 × 10^6 ft^2) = 2.232 inches. The flow ordinates are reduced by this factor, as shown in the table.

Time (hours)	Q (cfs)	Volume (cf)	Unit Hydrograph Q (cfs)
0-2	30	216,000	= 30/2.232 = 13.44
2-4	45	324,000	20.16
4-6	42	302,400	18.82
6-8	40	288,000	17.92
8-10	33	237,600	14.78
10-12	26	187,200	11.65
Total Volume		1,555,200	--

Problem

A watershed has the following data recorded during a storm event:
Volume of direct runoff = 1,500,000 cf
Watershed area = 861,079 sf (19.77 ac)
Peak discharge = 7800 cfs

Calculate a) the unit hydrograph peak discharge and b) the peak flow and design runoff volume (in cf) for a 3.2-inch net precipitation storm event (of same duration).

Solution
a) First, calculate the average precipitation, P_{ave}, for the watershed:
$$P_{ave} = \frac{Vol. \, of \, Direct \, Runoff}{Watershed \, Area} = \frac{1,500,000 \, ft^3}{861,079 \, ft^2} = 1.742 \, in$$
The peak unit discharge, $Q_{UH, \, peak}$ may then be calculated from:
$$Q_{UH,peak} = \frac{Q_{peak}}{P_{ave}} = \frac{7800 \, cfs}{1.742 \, in} = 4478 \frac{ft^3}{s*in}$$
b) The peak flow for 3.2-inch net precipitation is then:
$$Q_{3.2,peak} = Q_{UH,peak} P_{3.2,ave} = \frac{4478 \, ft^3}{s*in}(3.2 \, in) = 14,330 \, cfs$$
To calculate the design volume for a 3.2-inch storm, first calculate the unit hydrograph (P_{ave} = 1 in/in) total volume from:
$$V_{UH} = P_{ave} A = 1 \frac{in}{in} (861,079 \, ft^2) \left(\frac{1 \, ft}{12 \, in}\right) = 71,757 \, cf/in$$
The design volume for a 3.2-inch storm can be calculated from:
$$V_{3.2} = V_{UH} P_{3.2} = \left(71,757 \frac{cf}{in}\right)(3.2 \, in) = 229,622 \, cf$$

Intensity-duration-frequency curves

Intensity-duration-frequency (IDF) curves are curves shown on a graph of rainfall intensity against storm duration. Each curve on the graph represents the corresponding values of a certain storm frequency for the region being evaluated. The rainfall intensity can be quickly estimated based on the storm frequency and duration being used for project design. An example of graphed IDF curves is shown below:

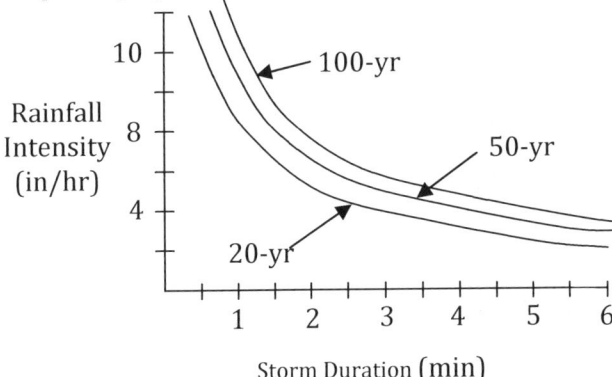

Problem

Use Horton's infiltration equation to calculate the rainfall that infiltrates the soil during a 7-hour storm in an area with the following characteristics:
Initial infiltration rate = 0.6 in/hr
Final infiltration rate = 0.3 in/hr
Horton's empirical constant, k = 0.5/hr

Solution
Horton's equation for infiltration of water into the soil is:
$$f = f_f + (f_o - f_f)e^{-kt}$$
Where f is equal to the infiltration rate (in/hr), f_f is the final infiltration rate, f_o is the initial infiltration rate, k is an empirically determined constant (1/hr), and t is time (hr). The infiltrated volume of rainfall can be determined by:
$$V_{inf} = \int_{t1}^{t2} f\, dt = \int_{t1}^{t2} (f_f + (f_o - f_f)e^{-kt})\, dt$$
Substituting the values of $f_f, f_o, t,$ and k into this equation yields:
$$V_{inf} = \int_0^7 (0.3 + (0.6 - 0.3)e^{-0.5t})\, dt$$
$$V_{inf} = 0.3 \int_0^7 (1 + e^{-0.5t})\, dt = 0.3 \left[t - \frac{1}{0.5}e^{-0.5t}\right]_0^7$$
$$V_{inf} = 0.3 \left[7 - \frac{1}{0.5}e^{-3.5} - 0 + \frac{1}{0.5}e^0\right] = 2.68 \text{ inches}$$

Problem

Runoff is collected from two areas A and B at point X. The storm intensity measured after 29 minutes is 2.7 inches per hour. Calculate the peak flow in cfs at point X using the rational method.

Area A
Area = 5 acres
Runoff coefficient = 0.68
Runoff travel time = 12 min

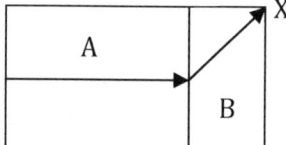

Area B
Area = 2.3 acres
Runoff coefficient = 0.41
Runoff travel time = 17 min

Solution
The calculation of peak flow, Q_p, using the rational method is shown below:
$$Q_p = CIA$$
Where C is the runoff coefficient, I is the storm intensity, and A is the watershed area in acres. The runoff coefficient for more than one area must be weighted by the contributing areas. For this example, the runoff coefficient is weighted by two contributing areas:
$$C = \frac{C_1 A_1 + C_2 A_2}{A_1 + A_2} = \frac{(0.68)(5\ ac) + (0.41)(2.3\ ac)}{5\ ac + 2.3\ ac} = 0.59$$
The peak flow is then:
$$Q_p = (0.59)\left(2.7 \frac{in}{hr}\right)(5\ ac + 2.3\ ac) = 11.63\ ac \times in/hr$$
$$Q_p = \left(11.63\ ac \times \frac{in}{hr}\right)\left(\frac{1.008 ft^3/s}{1\ ac \times in/hr}\right) = 11.72\ ft^3/s$$

Problem

Three watersheds tie separately into a pipeline. Watershed A is upstream of Watersheds B and C. The runoff travel time within Watersheds A, B, and C are 15 minutes, 18 minutes, and 20 minutes, respectively. The pipeline tie-in for Watershed A is 900 feet upstream of Watershed B's tie-in and 1440 feet upstream of Watershed C's tie-in. Assume a velocity of 6 ft/s, a slope of 0.008, and a Manning's roughness coefficient of 0.014 for the entire pipeline. Calculate the maximum time of concentration for the system. Also, calculate the pipeline design diameter (full flow) after Watershed C's tie-in for a total system flow of 22 cfs.

<u>Solution</u>
First calculate the pipeline travel time between tie-in points:
$$t_{A \to B} = \left(\frac{900\ ft}{6\ ft/s}\right)\left(\frac{1\ min}{60\ s}\right) = 2.5\ min$$
$$t_{B \to C} = \left(\frac{1440\ ft}{6\ ft/s}\right)\left(\frac{1\ min}{60\ s}\right) = 4\ min$$
Calculate the time it takes the water to travel through each watershed to Watershed C's tie-in:
$$t_{c,A \to C} = 15\ min + 2.5\ min + 4\ min = 21.5\ min$$
$$t_{c,B \to C} = 18\ min + 4\ min = 22\ min;\ t_{c,C} = 20\ min$$
The maximum time of concentration for the system is 22 min (through Watershed B to tie-in at C). The diameter, D, of the pipe can be calculated from:
$$D = 1.33 \left(\frac{nQ}{\sqrt{S}}\right)^{3/8}$$
Where n is Manning's roughness coefficient, Q is total flow, and S is the pipe slope in decimal format.
$$D = 1.33 \left(\frac{(0.014)(22\ cfs)}{\sqrt{(0.008)}}\right)^{3/8} = 2.11\ ft \approx 2.5\ ft$$

Problem

Use the SCS method to calculate the basin lag time for a wooded watershed having good cover with soil group B. The average slope of the watershed is 2.3% and the length of the main stream from the outlet to the divide is 100,000 feet.

<u>Solution</u>
The lag time, t_p, (in hours) is calculated from:
$$t_p = \frac{l^{0.8}(S+1)^{0.7}}{1900\ y^{0.5}}$$
Where l is the length of the main stream from the outlet to the divide (in feet), y is the average watershed slope (in percent), and S is potential abstraction (in inches). S is equal to:
$$S = \frac{1000}{CN} - 10$$
Where CN is the runoff curve number for the soil/land use (tabulated value). For a wooded area having good cover with soil group B, the curve number is equal to 55. The value of S is then:

$$S = \frac{1000}{55} - 10 = 8.18 \text{ in}$$

The lag time is then:
$$t_p = \frac{(100{,}000 \text{ ft})^{0.8}(8.18 \text{ in} + 1)^{0.7}}{1900 \, (2.3\%)^{0.5}} = 16.38 \text{ hr}$$

There are 16.38 hours between the centroid of the rainfall and the peak discharge for the watershed.

Problem

Discuss water velocities within and side slopes of erodible channels.

Solution

Velocities should be taken into consideration when designing channels made of erodible materials. If the velocity of the channel is too high for the channel material, the channel will erode over time. The maximum recommended velocities are dependent on the channel material and the components within the water being carried. For water carrying fine silts, recommended maximum velocities range from 2.5 ft/s for fine sand channels to 6.0 ft/s for shale channels (1.5 ft/s and 6.0 ft/s for clear water). The recommended channel side slopes depend on the erodible material the channel is made of. For example, if a channel's side slopes consist of firm rock, the walls could be vertical. For large channels consisting of firm earth, the channel wall slope (horizontal:vertical) is recommended to be no steeper than 1:1. And for erodible materials such as a sandy loam, the recommended slopes may be as mild as 3:1. In general, the side slope of an erodible channel should not exceed the channel material's natural angle of repose.

Runge-Kutta method

The *Runge-Kutta method* is an accurate numerical technique used to solve the continuity and storage equations with the goal of determining how the detention basin outflow and the basin storage vary with time for a known inflow hydrograph. The main equation for this method (from the continuity equation) is:

$$\frac{\Delta H}{\Delta t} = \frac{Q_{in}(t) - Q_{out}(H)}{A_r H}$$

Where $Q_{in}(t)$ is the detention basin inflow as a function of time, $Q_{out}(H)$ is the detention basin outflow as a function of basin head, and $A_r H$ is the surface area as a function of head. The Runge-Kutta method can be used with orders of accuracy ranging from first to fourth (k_1 to k_4). For fourth-order accuracy:

$$H_{n+1} = H_n + \frac{1}{6}[k_1 + 2k_2 + 2k_3 + k_4]\Delta t$$

$$k_1 = f(t_n, H_n), \quad k_2 = f\left(t_n + \frac{\Delta t}{2}, H_n + \frac{1}{2}k_1 \Delta t\right)$$

$$k_3 = f\left(t_n + \frac{\Delta t}{2}, H_n + \frac{1}{2}k_2 \Delta t\right)$$

$$k_4 = f(t_n + \Delta t, H_n + k_3 \Delta t)$$

Breakpoint chlorination process

Wastewater typically contains significant levels of ammonia nitrogen (NH_4^+ or NH_3). As chlorine is introduced in these types of waters, chloramines form. The type of chloramines formed depends on the pH of the water. Examples of chloramines are monochloramines (NH_2Cl), dichloramines ($NHCl_2$), and trichloramines (NCl_3). As additional chlorine is introduced, the chloramines are destroyed and nitrogen gas (N_2) and nitrous oxide (N_2O) form. With the continued application of chlorine, the chloramines continue to be converted and a phenomenon known as the breakpoint occurs. The breakpoint is the point at which all ammonia has been removed from the water. If additional chlorine is added following the breakpoint, free chlorine residuals form. This is known as *breakpoint chlorination*. The free chlorine residuals have a high disinfection capacity and are usually in the form of free chlorine (Cl_2), hypochlorous acid (HOCl), and hypochlorite ions.

Problem

Three 100-ft diameter final clarifiers are constructed in an activated sludge plant. Each clarifier has a side-water depth of 10 feet. Calculate the detention time and the overflow rate based on a plant design flow of 20 mgd. Also, calculate the solids loading on the final clarifiers if the plant's aeration tank operates at a mixed-liquor suspended-solids concentration of 5,000 mg/L and a recirculation ratio of 0.55.

Solution
First calculate the volume of all three clarifiers:
$$V = 3(\pi)(50\ ft)^2(10\ ft)\left(\frac{7.48\ gal}{1\ ft^3}\right) = 1.76 \times 10^6\ gal$$
The detention time is calculated from:
$$t_d = \frac{V}{Q_C} = \left(\frac{1.76 \times 10^6\ gal}{2 \times 10^7\ gal/day}\right)\left(\frac{24\ hr}{1\ day}\right) = 2.1\ hr$$
The overflow rate, OF, is the plant design flow divided by the total clarifier surface area:
$$OF = \frac{Q_C}{SA} = \frac{2 \times 10^7\ gal/day}{3(\pi)(50\ ft)^2} = 849\ gpd/ft^2$$
For a recirculation rate of 0.55, the flow from the aeration tank to the clarifiers is:
$1.55(20\ mgd) = 31\ mgd$
The solids loading, SL, is calculated from the following:
$$SL = \frac{Q_{AT}(MLSS)}{SA} = \frac{(31\ mgd)\left(5000\frac{mg}{L}\right)}{3(\pi)(50\ ft)^2}\left(\frac{8.34\ lb}{\left(\frac{mg}{L}\right)(1\ mgal)}\right)$$
$$SL = 55\ lb/(ft^2 \times d)$$

Problem

Provide and define the equations used to calculate BOD from a seeded and unseeded sample laboratory test.

Solution
The following equation is used for calculating BOD from a seeded laboratory test:
$$BOD = \frac{(D_1 - D_2) - (B_1 - B_2)f}{P}$$

Where D_1 is the concentration of dissolved oxygen, DO, (in mg/L) of the diluted seeded wastewater approximately 15 minutes after preparation, D_2 is the concentration of DO (in mg/L) after incubation, B_1 is the concentration of DO (in mg/L) of the diluted seed sample approximately 15 minutes after preparation, B_2 is the concentration of DO (in mg/L) of the seed sample after incubation, f is equal to the ratio of the seed volume in the seeded wastewater test to the seed volume in the BOD test on seed, and P is the decimal fraction of wastewater sample used (equal to the volume of wastewater divided by the volume of dilution water plus wastewater). For an unseeded sample, the equation for calculating BOD is simplified to:

$$BOD = \frac{(D_1 - D_2)}{P}$$

Nitrification-denitrification process

The *nitrification-denitrification process* is a two-step nitrogen-removing process, commonly used in wastewater treatment. The first step in the step is biological nitrification. Biological nitrification involves the oxidation of ammonia to nitrite and the conversion of nitrite to nitrate by bacteria. The bacteria responsible for nitrification are considerably sensitive and are susceptible to a variety of inhibitors such as excess concentrations of ammonia and/or nitrous acid. The second step of the nitrogen-removing process is denitrification, also a biological process. During dentrification, nitrate, through a series of reactions, is converted to nitrogen gas under anoxic conditions. Dissolved oxygen plays an important role in that its presence will suppress the denitrification process. Nitrate reduction involves the following reactions:
$NO_3^- \rightarrow NO_2^- \rightarrow NO \rightarrow N_2O \rightarrow N_2$
The compounds NO, N2O and N2 are gaseous products.

Problem

A BOD_5 test is performed. The sample volume for the test is 5 mL. The sample is diluted to full volume in 500 mL BOD bottles. The average initial dissolved oxygen (DO) concentration is 7.7 mg/L and after 5 days, the average measured DO concentration is 4.5 mg/L. The deoxygenation rate constant is 0.12 day^{-1}. Calculate the standard BOD and the ultimate carbonaceous BOD.

Solution
The equation for calculating BOD_5 is:
$$BOD_5 = \frac{DO_i - DO_f}{\dfrac{V_{sample}}{V_{sample} + V_{dilution}}}$$
Where DO_i is the initial dissolved oxygen (DO) concentration, DO_f is the DO concentration after 5 days,
V_{sample} is the sample volume, and $V_{dilution}$ is the diluted volume. For this example:
$$BOD_5 = \frac{7.7 \; mg/L - 4.5 \; mg/L}{\dfrac{5 \; mL}{500 \; mL}} = 320 \; mg/L$$
Since the value of BOD5 is above 300 mg/L, the wastewater is considered to be strong. The ultimate BOD is calculated from:
$$BOD_{ult} = \frac{BOD_t}{1 - 10^{-K_d t}}$$

Where K_d is the deoxygenation rate constant.
$$BOD_{ult} = \frac{320 \; mg/L}{1 - 10^{-(0.12 \; day^{-1})(5 \; days)}} = 427 \; mg/L$$

Problem

A wastewater treatment plant's primary clarifier receives 7 mgd. The clarifier is 100 feet in diameter and has a side water depth of 10 feet. Calculate the detention time, surface loading, and weir loading of this clarifier.

Solution
First calculate the surface area, A, of the primary clarifier: A = π(100 ft/2)² = 7854 ft². The volume, V, of the clarifier is then: V = 7854 ft²(10 ft) = 78,540 ft³. The detention time, t_d, for the primary clarifier is:
$$t_d = \frac{V}{Q} = \frac{78{,}540 \; ft^3 \left(24 \frac{hr}{day}\right)}{\left(\frac{7 \times 10^6 gal}{day}\right)\left(0.1337 \frac{ft^3}{gal}\right)} = 2.01 \; hr$$
The surface loading, SL, is:
$$SL = \frac{Q}{A} = \frac{\left(\frac{7 \times 10^6 gal}{day}\right)}{7854 \; ft^2} = 891 \; gal/(day \times ft^2)$$
The weir loading, WL, is:
$$WL = \frac{Q}{Length_{perimeter}} = \frac{\left(\frac{7 \times 10^6 gal}{day}\right)}{\pi(100 \; ft)} = 22{,}282 \frac{gal}{day \times ft}$$
All of these calculated values fall within design recommendations except for weir loading (typically between 10,000 and 20,000 gal/(day*ft) for primary clarifiers).

Problem

A wastewater treatment plant's single-stage high-rate trickling filter receives 1.5 mgd. The filter is 100 feet in diameter and has a recirculation ratio of 0.4. Calculate the hydraulic load of this trickling filter. Also, calculate the hydraulic load of the filter if it was a standard-rate filter instead of a high-rate filter.

Solution
The hydraulic load of a trickling filter is calculated from the following equation:
$$L_{hydraulic} = \frac{Q_w(1 + R)}{A}$$
Where Q_w is the flow rate of the influent to the trickling filter, R is the recirculation ratio, and A is the filter surface area. The filter surface area is:
$$A = \pi r^2 = \pi \left(\frac{100 \; ft}{2}\right)^2 = 7854 \; ft^2$$
The hydraulic load is then:
$$L_{hydraulic} = \frac{\left(\frac{1.5 \times 10^6 gal}{day}\right)(1 + 0.4)}{7854 \; ft^2} = 267 \frac{gal}{day \times ft^2}$$

A standard-rate filter has a recirculation ratio of zero. The hydraulic load for a standard-rate filter would then be:

$$L_{hydraulic} = \frac{\left(\frac{1.5 \times 10^6 \, gal}{day}\right)(1+0)}{7854 \, ft^2} = 191 \frac{gal}{day \times ft^2}$$

Problem

Define the mass-action formula for a reaction in true equilibrium. Also, determine the equilibrium constants for the two following reactions:

$2SO_{2(g)} + O_{2(g)} \rightleftharpoons 2SO_{3(g)}$

$H_2CO_3 \rightleftharpoons H^+ + HCO_3^-$

Solution
Consider the following reaction in true equilibrium:
$aA + bB \rightleftharpoons cC + dD$
Where A and B are reactants, C and D are products, and a, b, c, and d are the number of moles of each reactant and product. The equilibrium constant for the reaction, K, is determined from the mass-action formula:

$$K = \frac{[C]^c[D]^d}{[A]^a[B]^b}$$

Where [C], [D], [A], and [B] are the molar concentrations of the products and reactants. The equilibrium constant for the equilibrium reactions given are:

$$K = \frac{\left[SO_{3(g)}\right]^2}{\left[SO_{2(g)}\right]^2 [O_{2(g)}]}$$

$$K = \frac{[H^+][HCO_3^-]}{[H_2CO_3]}$$

Problem

Water flows at a rate of 2500 cubic feet per minute through a rectangular basin that is 10 feet deep, 20 feet wide, and 100 feet long. Determine if a particle with a settling velocity of 1 foot per minute will settle in the basin.

Solution
A particle will settle if the time it takes for the particle to settle is less than the particle's detention time within the basin. The equation for calculating minimum settling time for a particle is dependent on the tank depth, h and the particle settling velocity, v_s:

$t_{settling} = h/v_s$

For this example, the minimum settling time is:

$$t_{settling} = \frac{10 \, ft}{1 \, ft/min} = 10 \, min$$

The equation to calculate the particle's detention time in the basin is dependent on basin volume, V_{basin}, and flow, Q:

$$t_{detention} = \frac{V_{basin}}{Q}$$

$$t_{detention} = \frac{(10\ ft)(20\ ft)(100\ ft)}{2500\ ft^3/min} = 8\ min$$

The particle will not settle since $t_{settling}$ is greater than $t_{detention}$ (10 min > 8 min).

Problem

A flocculation tank within a water treatment plant is 50 feet long and 75 feet wide. The water depth within the tank is 15 feet. Baffles separate the compartments that house four horizontal shafts of paddle flocculators. Each of the four flocculators have four arms with two blades at radii 2.5 feet and 5 feet from the shaft to the center of the 0.8 ft boards. The length of the paddle boards across the tank is 40 feet total. Assuming a ratio of water velocity to paddle velocity of 0.4, a drag coefficient of 1.8 for flat plates, a paddle rotational speed of 0.05 rev/s, and a water density of 1.938 lb*s/ft⁴, calculate the power dissipated by the paddle flocculators.

Solution
The power, P, dissipated in a flocculation tank by paddle flocculators with symmetrical paddles arms with blades at two radii can be calculated from:
$$P = \frac{n}{2} C_d A \rho (1-k)^3 (2\pi N)^3 (r_1^3 + r_2^3)$$
Where n is the number of symmetrical paddle arms, C_d is the drag coefficient, A is the area of the paddles, ρ is the density of water, k is the ratio of water velocity to paddle velocity, N is the rotational speed of the paddles, and r_1 and r_2 are the distances from the shaft to the paddle centers (radii). The number of paddle arms is 4 units x 4 arms = 16 arms. The area of the paddles is 40 feet x 0.8 ft = 32 ft².

The power dissipated is then:
$$P = \frac{16}{2}(1.8)(32)(1.938)(1-0.4)^3(2\pi(0.05))^3(2.5^3 + 5^3)$$

$$P = 841\ ft \times lb/s$$

Problem

Provide the reaction for alum reacting with natural alkalinity in water. Determine how many alkalinity ions react for every molecule of alum introduced. Also, calculate the amount of alum (in mg/L) required to counteract a concentration of 2.8 mg/L of natural alkalinity in a water treatment plant.

Solution
The reaction between alum and natural alkalinity in water is:
$Al_2(SO_4)_3 \cdot XH_2O + 3Ca(HCO_3)_2 \rightarrow$
$2Al(OH)_3 \downarrow + 3CaSO_4 + 6CO_2 + XH_2O$
Where $Al_2(SO_4)_3$ is alum and HCO_3^- is an alkalinity ion. From the above balanced reaction, each molecule of alum reacts with six (6) alkalinity ions. The molecular weight of alum is:
$MW_{Alum} = (2 \times 27) + (3 \times 32) + (12 \times 16) = 342$
The molecular weight of six alkalinity ions is:
$MW_{Alkalinity} = (6 \times 1) + (6 \times 12) + (18 \times 16) = 366$
The amount of alum needed to counteract 2.8 mg/L of natural alkalinity in a water treatment plant is calculated from:

$$Alum_{mg/L} = \frac{Alkalinity_{mg/L} \times MW_{Alum}}{MW_{Alkalinity}}$$

$$Alum_{mg/L} = \frac{\left(2.8 \frac{mg}{L}\right)(342)}{366} = 2.62 \frac{mg}{L} \, alum$$

Problem

Influent to a 10 mgd water treatment plant contains natural alkalinity in the form of HCO_3^-. Assuming a purity of 100%, calculate the feed rate if alum is dosed at 15 mg/L.

Solution
The dose equation for calculating feed rate, F, of a compound given its purity, P and its availability G is shown below for both SI and U.S. units:

$$F_{kg/day} = \frac{D_{mg/L} Q_{mL/day}}{PG} \quad (Units: SI)$$

$$F_{lbm/day} = \frac{D_{mg/L} Q_{MGD} \left(8.345 \frac{lbm \times L}{mg \times MG}\right)}{PG} \quad (Units: U.S.)$$

Where D is the dose and Q is the flow rate.
For the information given, the feed rate is:

$$F_{lbm/day} = \frac{\left(15 \frac{mg}{L}\right)(10 \, MGD)\left(8.345 \frac{lbm \times L}{mg \times MG}\right)}{(1.0)(1.0)}$$

$$F_{lbm/day} = 1252 \, lbm/day$$

Problem
Provide equations for approximating dry sludge production rates for the following sludges: dry alum, dry ferric sulfate, and dry ferric chloride.

Solution
The production rate for dry alum sludge in pounds per million gallons of water treated can be approximated from:
$$Sludge_{alum} = [(Dose_{alum})(0.26)(8.34)] + [(T)(R)(8.34)]$$
Where $Dose_{alum}$ is in mg/L, T is the raw water turbidity in ntu, and R is the ratio between total suspended solids in mg/L and turbidity in ntu.
The production rate for dry ferric sulfate sludge in pounds per million gallons of water treated can be approximated from:
$$Sludge_{FS} = [(Dose_{FS})(0.54)(8.34)] + [(T)(R)(8.34)]$$
Where $Dose_{FS}$ is in mg/L.
The production rate for dry ferric sulfate sludge in pounds per million gallons of water treated can be approximated from:
$$Sludge_{FC} = [(Dose_{FC})(0.66)(8.34)] + [(T)(R)(8.34)]$$
Where $Dose_{FC}$ is in mg/L.
The ratio between suspended solids and turbidity is typically within the range of 1.0 to 2.0.

Secret Key #1 - Time is Your Greatest Enemy

Pace Yourself

Wear a watch. At the beginning of the test, check the time (or start a chronometer on your watch to count the minutes), and check the time after every few questions to make sure you are "on schedule."

If you are forced to speed up, do it efficiently. Usually one or more answer choices can be eliminated without too much difficulty. Above all, don't panic. Don't speed up and just begin guessing at random choices. By pacing yourself, and continually monitoring your progress against your watch, you will always know exactly how far ahead or behind you are with your available time. If you find that you are one minute behind on the test, don't skip one question without spending any time on it, just to catch back up. Take 15 fewer seconds on the next four questions, and after four questions you'll have caught back up. Once you catch back up, you can continue working each problem at your normal pace.

Furthermore, don't dwell on the problems that you were rushed on. If a problem was taking up too much time and you made a hurried guess, it must be difficult. The difficult questions are the ones you are most likely to miss anyway, so it isn't a big loss. It is better to end with more time than you need than to run out of time.

Lastly, sometimes it is beneficial to slow down if you are constantly getting ahead of time. You are always more likely to catch a careless mistake by working more slowly than quickly, and among very high-scoring test takers (those who are likely to have lots of time left over), careless errors affect the score more than mastery of material.

Secret Key #2 - Guessing is not Guesswork

You probably know that guessing is a good idea. Unlike other standardized tests, there is no penalty for getting a wrong answer. Even if you have no idea about a question, you still have a 20-25% chance of getting it right.

Most test takers do not understand the impact that proper guessing can have on their score. Unless you score extremely high, guessing will significantly contribute to your final score.

Monkeys Take the Test

What most test takers don't realize is that to insure that 20-25% chance, you have to guess randomly. If you put 20 monkeys in a room to take this test, assuming they answered once per question and behaved themselves, on average they would get 20-25% of the questions correct. Put 20 test takers in the room, and the average will be much lower among guessed questions. Why?

1. The test writers intentionally write deceptive answer choices that "look" right. A test taker has no idea about a question, so he picks the "best looking" answer, which is often wrong. The monkey has no idea what looks good and what doesn't, so it will consistently be right about 20-25% of the time.
2. Test takers will eliminate answer choices from the guessing pool based on a hunch or intuition. Simple but correct answers often get excluded, leaving a 0% chance of being correct. The monkey has no clue, and often gets lucky with the best choice.

This is why the process of elimination endorsed by most test courses is flawed and detrimental to your performance. Test takers don't guess; they make an ignorant stab in the dark that is usually worse than random.

$5 Challenge

Let me introduce one of the most valuable ideas of this course—the $5 challenge:

You only mark your "best guess" if you are willing to bet $5 on it.
You only eliminate choices from guessing if you are willing to bet $5 on it.

Why $5? Five dollars is an amount of money that is small yet not insignificant, and can really add up fast (20 questions could cost you $100). Likewise, each answer choice on one question of the test will have a small impact on your overall score, but it can really add up to a lot of points in the end.

The process of elimination IS valuable. The following shows your chance of guessing it right:

If you eliminate wrong answer choices until only this many remain:	Chance of getting it correct:
1	100%
2	50%
3	33%

However, if you accidentally eliminate the right answer or go on a hunch for an incorrect answer, your chances drop dramatically—to 0%. By guessing among all the answer choices, you are GUARANTEED to have a shot at the right answer.

That's why the $5 test is so valuable. If you give up the advantage and safety of a pure guess, it had better be worth the risk.

What we still haven't covered is how to be sure that whatever guess you make is truly random. Here's the easiest way:

Always pick the first answer choice among those remaining.

Such a technique means that you have decided, **before you see a single test question**, exactly how you are going to guess, and since the order of choices tells you nothing about which one is correct, this guessing technique is perfectly random.

This section is not meant to scare you away from making educated guesses or eliminating choices; you just need to define when a choice is worth eliminating. The $5 test, along with a pre-defined random guessing strategy, is the best way to make sure you reap all of the benefits of guessing.

Secret Key #3 - Practice Smarter, Not Harder

Many test takers delay the test preparation process because they dread the awful amounts of practice time they think necessary to succeed on the test. We have refined an effective method that will take you only a fraction of the time.

There are a number of "obstacles" in the path to success. Among these are answering questions, finishing in time, and mastering test-taking strategies. All must be executed on the day of the test at peak performance, or your score will suffer. The test is a mental marathon that has a large impact on your future.

Just like a marathon runner, it is important to work your way up to the full challenge. So first you just worry about questions, and then time, and finally strategy:

Success Strategy

1. Find a good source for practice tests.
2. If you are willing to make a larger time investment, consider using more than one study guide. Often the different approaches of multiple authors will help you "get" difficult concepts.
3. Take a practice test with no time constraints, with all study helps, "open book." Take your time with questions and focus on applying strategies.
4. Take a practice test with time constraints, with all guides, "open book."
5. Take a final practice test without open material and with time limits.

If you have time to take more practice tests, just repeat step 5. By gradually exposing yourself to the full rigors of the test environment, you will condition your mind to the stress of test day and maximize your success.

Secret Key #4 - Prepare, Don't Procrastinate

Let me state an obvious fact: if you take the test three times, you will probably get three different scores. This is due to the way you feel on test day, the level of preparedness you have, and the version of the test you see. Despite the test writers' claims to the contrary, some versions of the test WILL be easier for you than others.

Since your future depends so much on your score, you should maximize your chances of success. In order to maximize the likelihood of success, you've got to prepare in advance. This means taking practice tests and spending time learning the information and test taking strategies you will need to succeed.

Never go take the actual test as a "practice" test, expecting that you can just take it again if you need to. Take all the practice tests you can on your own, but when you go to take the official test, be prepared, be focused, and do your best the first time!

Secret Key #5 - Test Yourself

Everyone knows that time is money. There is no need to spend too much of your time or too little of your time preparing for the test. You should only spend as much of your precious time preparing as is necessary for you to get the score you need.

Once you have taken a practice test under real conditions of time constraints, then you will know if you are ready for the test or not.

If you have scored extremely high the first time that you take the practice test, then there is not much point in spending countless hours studying. You are already there.

Benchmark your abilities by retaking practice tests and seeing how much you have improved. Once you consistently score high enough to guarantee success, then you are ready.

If you have scored well below where you need, then knuckle down and begin studying in earnest. Check your improvement regularly through the use of practice tests under real conditions. Above all, don't worry, panic, or give up. The key is perseverance!

Then, when you go to take the test, remain confident and remember how well you did on the practice tests. If you can score high enough on a practice test, then you can do the same on the real thing.

General Strategies

The most important thing you can do is to ignore your fears and jump into the test immediately. Do not be overwhelmed by any strange-sounding terms. You have to jump into the test like jumping into a pool—all at once is the easiest way.

Make Predictions

As you read and understand the question, try to guess what the answer will be. Remember that several of the answer choices are wrong, and once you begin reading them, your mind will immediately become cluttered with answer choices designed to throw you off. Your mind is typically the most focused immediately after you have read the question and digested its contents. If you can, try to predict what the correct answer will be. You may be surprised at what you can predict.

Quickly scan the choices and see if your prediction is in the listed answer choices. If it is, then you can be quite confident that you have the right answer. It still won't hurt to check the other answer choices, but most of the time, you've got it!

Answer the Question

It may seem obvious to only pick answer choices that answer the question, but the test writers can create some excellent answer choices that are wrong. Don't pick an answer just because it sounds right, or you believe it to be true. It MUST answer the question. Once you've made your selection, always go back and check it against the question and make sure that you didn't misread the question and that the answer choice does answer the question posed.

Benchmark

After you read the first answer choice, decide if you think it sounds correct or not. If it doesn't, move on to the next answer choice. If it does, mentally mark that answer choice. This doesn't mean that you've definitely selected it as your answer choice, it just means that it's the best you've seen thus far. Go ahead and read the next choice. If the next choice is worse than the one you've already selected, keep going to the next answer choice. If the next choice is better than the choice you've already selected, mentally mark the new answer choice as your best guess.

The first answer choice that you select becomes your standard. Every other answer choice must be benchmarked against that standard. That choice is correct until proven otherwise by another answer choice beating it out. Once you've decided that no other answer choice seems as good, do one final check to ensure that your answer choice answers the question posed.

Valid Information

Don't discount any of the information provided in the question. Every piece of information may be necessary to determine the correct answer. None of the information in the question is there to throw you off (while the answer choices will certainly have information to throw you off). If two seemingly unrelated topics are discussed, don't ignore either. You can be confident there is a relationship, or it wouldn't be included in the question, and you are probably going to have to determine what is that relationship to find the answer.

Avoid "Fact Traps"

Don't get distracted by a choice that is factually true. Your search is for the answer that answers the question. Stay focused and don't fall for an answer that is true but irrelevant. Always go back to the question and make sure you're choosing an answer that actually answers the question and is not just a true statement. An answer can be factually correct, but it MUST answer the question asked. Additionally, two answers can both be seemingly correct, so be sure to read all of the answer choices, and make sure that you get the one that BEST answers the question.

Milk the Question

Some of the questions may throw you completely off. They might deal with a subject you have not been exposed to, or one that you haven't reviewed in years. While your lack of knowledge about the subject will be a hindrance, the question itself can give you many clues that will help you find the correct answer. Read the question carefully and look for clues. Watch particularly for adjectives and nouns describing difficult terms or words that you don't recognize. Regardless of whether you completely understand a word or not, replacing it with a synonym, either provided or one you more familiar with, may help you to understand what the questions are asking. Rather than wracking your mind about specific detailed information concerning a difficult term or word, try to use mental substitutes that are easier to understand.

The Trap of Familiarity

Don't just choose a word because you recognize it. On difficult questions, you may not recognize a number of words in the answer choices. The test writers don't put "make-believe" words on the test, so don't think that just because you only recognize all the words in one answer choice that that answer choice must be correct. If you only recognize words in one answer choice, then focus on that one. Is it correct? Try your best to determine if it is correct. If it is, that's great. If not, eliminate it. Each word and answer choice you eliminate increases your chances of getting the question correct, even if you then have to guess among the unfamiliar choices.

Eliminate Answers

Eliminate choices as soon as you realize they are wrong. But be careful! Make sure you consider all of the possible answer choices. Just because one appears right, doesn't mean that the next one won't be even better! The test writers will usually put more than one good answer choice for every question, so read all of them. Don't worry if you are stuck between two that seem right. By getting down to just two remaining possible choices, your odds are now 50/50. Rather than wasting too much time, play the odds. You are guessing, but guessing wisely because you've been able to knock out some of the answer choices that you know are wrong. If you are eliminating choices and realize that the last answer choice you are left with is also obviously wrong, don't panic. Start over and consider each choice again. There may easily be something that you missed the first time and will realize on the second pass.

Tough Questions

If you are stumped on a problem or it appears too hard or too difficult, don't waste time. Move on! Remember though, if you can quickly check for obviously incorrect answer choices, your chances of guessing correctly are greatly improved. Before you completely give up, at least try to knock out a couple of possible answers. Eliminate what you can and then guess at the remaining answer choices before moving on.

Brainstorm

If you get stuck on a difficult question, spend a few seconds quickly brainstorming. Run through the complete list of possible answer choices. Look at each choice and ask yourself, "Could this answer the question satisfactorily?" Go through each answer choice and consider it independently of the others. By systematically going through all possibilities, you may find something that you would otherwise overlook. Remember though that when you get stuck, it's important to try to keep moving.

Read Carefully

Understand the problem. Read the question and answer choices carefully. Don't miss the question because you misread the terms. You have plenty of time to read each question thoroughly and make sure you understand what is being asked. Yet a happy medium must be attained, so don't waste too much time. You must read carefully, but efficiently.

Face Value

When in doubt, use common sense. Always accept the situation in the problem at face value. Don't read too much into it. These problems will not require you to make huge leaps of logic. The test writers aren't trying to throw you off with a cheap trick. If you have to go beyond creativity and make a leap of logic in order to have an answer choice answer the question, then you should look at the other answer choices. Don't overcomplicate the problem by creating theoretical relationships or explanations that will warp time or space. These are normal problems rooted in reality. It's just that the applicable relationship or explanation may not be readily apparent and you have to figure things out. Use your common sense to interpret anything that isn't clear.

Prefixes

If you're having trouble with a word in the question or answer choices, try dissecting it. Take advantage of every clue that the word might include. Prefixes and suffixes can be a huge help. Usually they allow you to determine a basic meaning. Pre- means before, post- means after, pro - is positive, de- is negative. From these prefixes and suffixes, you can get an idea of the general meaning of the word and try to put it into context. Beware though of any traps. Just because con- is the opposite of pro-, doesn't necessarily mean congress is the opposite of progress!

Hedge Phrases

Watch out for critical hedge phrases, led off with words such as "likely," "may," "can," "sometimes," "often," "almost," "mostly," "usually," "generally," "rarely," and "sometimes." Question writers insert these hedge phrases to cover every possibility. Often an answer choice will be wrong simply because it leaves no room for exception. Unless the situation calls for them, avoid answer choices that have definitive words like "exactly," and "always."

Switchback Words

Stay alert for "switchbacks." These are the words and phrases frequently used to alert you to shifts in thought. The most common switchback word is "but." Others include "although," "however," "nevertheless," "on the other hand," "even though," "while," "in spite of," "despite," and "regardless of."

New Information

Correct answer choices will rarely have completely new information included. Answer choices typically are straightforward reflections of the material asked about and will directly relate to the question. If a new piece of information is included in an answer choice that doesn't even seem to relate to the topic being asked about, then that answer choice is likely incorrect. All of the information needed to answer the question is usually provided for you in the question. You should not have to make guesses that are unsupported or choose answer choices that require unknown information that cannot be reasoned from what is given.

Time Management

On technical questions, don't get lost on the technical terms. Don't spend too much time on any one question. If you don't know what a term means, then odds are you aren't going to get much further since you don't have a dictionary. You should be able to immediately recognize whether or not you know a term. If you don't, work with the other clues that you have—the other answer choices and terms provided—but don't waste too much time trying to figure out a difficult term that you don't know.

Contextual Clues

Look for contextual clues. An answer can be right but not the correct answer. The contextual clues will help you find the answer that is most right and is correct. Understand the context in which a phrase or statement is made. This will help you make important distinctions.

Don't Panic

Panicking will not answer any questions for you; therefore, it isn't helpful. When you first see the question, if your mind goes blank, take a deep breath. Force yourself to mechanically go through the steps of solving the problem using the strategies you've learned.

Pace Yourself

Don't get clock fever. It's easy to be overwhelmed when you're looking at a page full of questions, your mind is full of random thoughts and feeling confused, and the clock is ticking down faster than you would like. Calm down and maintain the pace that you have set for yourself. As long as you are on track by monitoring your pace, you are guaranteed to have enough time for yourself. When you get to the last few minutes of the test, it may seem like you won't have enough time left, but if you only have as many questions as you should have left at that point, then you're right on track!

Answer Selection

The best way to pick an answer choice is to eliminate all of those that are wrong, until only one is left and confirm that is the correct answer. Sometimes though, an answer choice may immediately look right. Be careful! Take a second to make sure that the other choices are not equally obvious. Don't make a hasty mistake. There are only two times that you should stop before checking other answers. First is when you are positive that the answer choice you have selected is correct. Second is when time is almost out and you have to make a quick guess!

Check Your Work

Since you will probably not know every term listed and the answer to every question, it is important that you get credit for the ones that you do know. Don't miss any questions through careless mistakes. If at all possible, try to take a second to look back over your answer selection and make sure you've selected the correct answer choice and haven't made a costly careless mistake (such as marking an answer choice that you didn't mean to mark). The time it takes for this quick double check should more than pay for itself in caught mistakes.

Beware of Directly Quoted Answers

Sometimes an answer choice will repeat word for word a portion of the question or reference section. However, beware of such exact duplication. It may be a trap! More than likely, the correct choice will paraphrase or summarize a point, rather than being exactly the same wording.

Slang

Scientific sounding answers are better than slang ones. An answer choice that begins "To compare the outcomes…" is much more likely to be correct than one that begins "Because some people insisted…"

Extreme Statements

Avoid wild answers that throw out highly controversial ideas that are proclaimed as established fact. An answer choice that states the "process should used in certain situations, if…" is much more likely to be correct than one that states the "process should be discontinued completely." The first is a calm rational statement and doesn't even make a definitive, uncompromising stance, using a hedge word "if" to provide wiggle room, whereas the second choice is a radical idea and far more extreme.

Answer Choice Families

When you have two or more answer choices that are direct opposites or parallels, one of them is usually the correct answer. For instance, if one answer choice states "x increases" and another answer choice states "x decreases" or "y increases," then those two or three answer choices are very similar in construction and fall into the same family of answer choices. A family of answer choices consists of two or three answer choices, very similar in construction, but often with directly opposite meanings. Usually the correct answer choice will be in that family of answer choices. The "odd man out" or answer choice that doesn't seem to fit the parallel construction of the other answer choices is more likely to be incorrect.

Special Report: How to Overcome Test Anxiety

The very nature of tests caters to some level of anxiety, nervousness, or tension, just as we feel for any important event that occurs in our lives. A little bit of anxiety or nervousness can be a good thing. It helps us with motivation, and makes achievement just that much sweeter. However, too much anxiety can be a problem, especially if it hinders our ability to function and perform.

"Test anxiety," is the term that refers to the emotional reactions that some test-takers experience when faced with a test or exam. Having a fear of testing and exams is based upon a rational fear, since the test-taker's performance can shape the course of an academic career. Nevertheless, experiencing excessive fear of examinations will only interfere with the test-taker's ability to perform and chance to be successful.

There are a large variety of causes that can contribute to the development and sensation of test anxiety. These include, but are not limited to, lack of preparation and worrying about issues surrounding the test.

Lack of Preparation

Lack of preparation can be identified by the following behaviors or situations:

Not scheduling enough time to study, and therefore cramming the night before the test or exam
Managing time poorly, to create the sensation that there is not enough time to do everything
Failing to organize the text information in advance, so that the study material consists of the entire text and not simply the pertinent information
Poor overall studying habits

Worrying, on the other hand, can be related to both the test taker, or many other factors around him/her that will be affected by the results of the test. These include worrying about:

Previous performances on similar exams, or exams in general
How friends and other students are achieving
The negative consequences that will result from a poor grade or failure

There are three primary elements to test anxiety. Physical components, which involve the same typical bodily reactions as those to acute anxiety (to be discussed below). Emotional factors have to do with fear or panic. Mental or cognitive issues concerning attention spans and memory abilities.

Physical Signals

There are many different symptoms of test anxiety, and these are not limited to mental and emotional strain. Frequently there are a range of physical signals that will let a test taker know that he/she is suffering from test anxiety. These bodily changes can include the following:

Perspiring
Sweaty palms
Wet, trembling hands
Nausea
Dry mouth
A knot in the stomach
Headache
Faintness
Muscle tension
Aching shoulders, back and neck
Rapid heart beat
Feeling too hot/cold

To recognize the sensation of test anxiety, a test-taker should monitor him/herself for the following sensations:

The physical distress symptoms as listed above
Emotional sensitivity, expressing emotional feelings such as the need to cry or laugh too much, or a sensation of anger or helplessness
A decreased ability to think, causing the test-taker to blank out or have racing thoughts that are hard to organize or control.

Though most students will feel some level of anxiety when faced with a test or exam, the majority can cope with that anxiety and maintain it at a manageable level. However, those who cannot are faced with a very real and very serious condition, which can and should be controlled for the immeasurable benefit of this sufferer.

Naturally, these sensations lead to negative results for the testing experience. The most common effects of test anxiety have to do with nervousness and mental blocking.

Nervousness

Nervousness can appear in several different levels:

The test-taker's difficulty, or even inability to read and understand the questions on the test
The difficulty or inability to organize thoughts to a coherent form
The difficulty or inability to recall key words and concepts relating to the testing questions (especially essays)
The receipt of poor grades on a test, though the test material was well known by the test taker

Conversely, a person may also experience mental blocking, which involves:

Blanking out on test questions
Only remembering the correct answers to the questions when the test has already finished.

Fortunately for test anxiety sufferers, beating these feelings, to a large degree, has to do with proper preparation. When a test taker has a feeling of preparedness, then anxiety will be dramatically lessened.

The first step to resolving anxiety issues is to distinguish which of the two types of anxiety are being suffered. If the anxiety is a direct result of a lack of preparation, this should be considered a normal reaction, and the anxiety level (as opposed to the test results) shouldn't be anything to worry about. However, if, when adequately prepared, the test-taker still panics, blanks out, or seems to overreact, this is not a fully rational reaction. While this can be considered normal too, there are many ways to combat and overcome these effects.

Remember that anxiety cannot be entirely eliminated, however, there are ways to minimize it, to make the anxiety easier to manage. Preparation is one of the best ways to minimize test anxiety. Therefore the following techniques are wise in order to best fight off any anxiety that may want to build.

To begin with, try to avoid cramming before a test, whenever it is possible. By trying to memorize an entire term's worth of information in one day, you'll be shocking your system, and not giving yourself a very good chance to absorb the information. This is an easy path to anxiety, so for those who suffer from test anxiety, cramming should not even be considered an option.

Instead of cramming, work throughout the semester to combine all of the material which is presented throughout the semester, and work on it gradually as the course goes by, making sure to master the main concepts first, leaving minor details for a week or so before the test.

To study for the upcoming exam, be sure to pose questions that may be on the examination, to gauge the ability to answer them by integrating the ideas from your texts, notes and lectures, as well as any supplementary readings.

If it is truly impossible to cover all of the information that was covered in that particular term, concentrate on the most important portions, that can be covered very well. Learn these concepts as best as possible, so that when the test comes, a goal can be made to use these concepts as presentations of your knowledge.

In addition to study habits, changes in attitude are critical to beating a struggle with test anxiety. In fact, an improvement of the perspective over the entire test-taking experience can actually help a test taker to enjoy studying and therefore improve the overall experience. Be certain not to overemphasize the significance of the grade - know that the result of the test is neither a reflection of self worth, nor is it a measure of intelligence; one grade will not predict a person's future success.

To improve an overall testing outlook, the following steps should be tried:

Keeping in mind that the most reasonable expectation for taking a test is to expect to try to demonstrate as much of what you know as you possibly can.
Reminding ourselves that a test is only one test; this is not the only one, and there will be others.
The thought of thinking of oneself in an irrational, all-or-nothing term should be avoided at all costs.
A reward should be designated for after the test, so there's something to look forward to. Whether it be going to a movie, going out to eat, or simply visiting friends, schedule it in advance, and do it no matter what result is expected on the exam.

Test-takers should also keep in mind that the basics are some of the most important things, even beyond anti-anxiety techniques and studying. Never neglect the basic social, emotional and biological needs, in order to try to absorb information. In order to best achieve, these three factors must be held as just as important as the studying itself.

Study Steps

Remember the following important steps for studying:

Maintain healthy nutrition and exercise habits. Continue both your recreational activities and social pass times. These both contribute to your physical and emotional well being.
Be certain to get a good amount of sleep, especially the night before the test, because when you're overtired you are not able to perform to the best of your best ability.
Keep the studying pace to a moderate level by taking breaks when they are needed, and varying the work whenever possible, to keep the mind fresh instead of getting bored.
When enough studying has been done that all the material that can be learned has been learned, and the test taker is prepared for the test, stop studying and do something relaxing such as listening to music, watching a movie, or taking a warm bubble bath.

There are also many other techniques to minimize the uneasiness or apprehension that is experienced along with test anxiety before, during, or even after the examination. In fact, there are a great deal of things that can be done to stop anxiety from interfering with lifestyle and performance. Again, remember that anxiety will not be eliminated entirely, and it shouldn't be. Otherwise that "up" feeling for exams would not exist, and most of us depend on that sensation to perform better than usual. However, this anxiety has to be at a level that is manageable.

Of course, as we have just discussed, being prepared for the exam is half the battle right away. Attending all classes, finding out what knowledge will be expected on the exam, and knowing the exam schedules are easy steps to lowering anxiety. Keeping up with work will remove the need to cram, and efficient study habits will eliminate wasted time. Studying should be done in an ideal location for concentration, so that it is simple to become interested in the material and give it complete attention. A method such as SQ3R (Survey, Question, Read, Recite, Review) is a wonderful key to follow to make sure that the study habits are as effective as possible, especially in the case of learning from a textbook. Flashcards are great techniques for memorization. Learning to take good notes will mean that notes will be full of useful information, so that less sifting will need to be done to seek out what is pertinent for studying. Reviewing notes after class and then again on occasion will keep the information fresh in the mind. From notes that have been taken summary sheets and outlines can be made for simpler reviewing.

A study group can also be a very motivational and helpful place to study, as there will be a sharing of ideas, all of the minds can work together, to make sure that everyone understands, and the studying will be made more interesting because it will be a social occasion.

Basically, though, as long as the test-taker remains organized and self confident, with efficient study habits, less time will need to be spent studying, and higher grades will be achieved.

To become self confident, there are many useful steps. The first of these is "self talk." It has been shown through extensive research, that self-talk for students who suffer from test anxiety, should be well monitored, in order to make sure that it contributes to self confidence as opposed to sinking the student. Frequently the self talk of test-anxious students is negative or self-defeating, thinking that everyone else is smarter and faster, that they always mess up, and that if they don't do well, they'll fail the entire course. It is important to decreasing anxiety that awareness is made of self talk. Try writing any negative self thoughts and then disputing them with a positive statement instead. Begin self-encouragement as though it was a friend speaking. Repeat positive statements to help reprogram the mind to believing in successes instead of failures.

Helpful Techniques

Other extremely helpful techniques include:

Self-visualization of doing well and reaching goals
While aiming for an "A" level of understanding, don't try to "overprotect" by setting your expectations lower. This will only convince the mind to stop studying in order to meet the lower expectations.
Don't make comparisons with the results or habits of other students. These are individual factors, and different things work for different people, causing different results.
Strive to become an expert in learning what works well, and what can be done in order to improve. Consider collecting this data in a journal.
Create rewards for after studying instead of doing things before studying that will only turn into avoidance behaviors.
Make a practice of relaxing - by using methods such as progressive relaxation, self-hypnosis, guided imagery, etc - in order to make relaxation an automatic sensation.
Work on creating a state of relaxed concentration so that concentrating will take on the focus of the mind, so that none will be wasted on worrying.
Take good care of the physical self by eating well and getting enough sleep.
Plan in time for exercise and stick to this plan.

Beyond these techniques, there are other methods to be used before, during and after the test that will help the test-taker perform well in addition to overcoming anxiety.

Before the exam comes the academic preparation. This involves establishing a study schedule and beginning at least one week before the actual date of the test. By doing this, the anxiety of not having enough time to study for the test will be automatically eliminated. Moreover, this will make the studying a much more effective experience, ensuring that the learning will be an easier process. This relieves much undue pressure on the test-taker.

Summary sheets, note cards, and flash cards with the main concepts and examples of these main concepts should be prepared in advance of the actual studying time. A topic should never be eliminated from this process. By omitting a topic because it isn't expected to be on the test is only setting up the test-taker for anxiety should it actually appear on the exam. Utilize the course syllabus for laying out the topics that should be studied. Carefully go over the notes that were made in class, paying special attention to any of the issues that the professor took special care to emphasize while lecturing in class. In the textbooks, use the chapter review, or if possible, the chapter tests, to begin your review.

It may even be possible to ask the instructor what information will be covered on the exam, or what the format of the exam will be (for example, multiple choice, essay, free form, true-false). Additionally, see if it is possible to find out how many questions will be on the test. If a review sheet or sample test has been offered by the professor, make good use of it, above anything else, for the preparation for the test. Another great resource for getting to know the examination is reviewing tests from previous semesters. Use these tests to review, and aim to achieve a 100% score on each of the possible topics. With a few exceptions, the goal that you set for yourself is the highest one that you will reach.

Take all of the questions that were assigned as homework, and rework them to any other possible course material. The more problems reworked, the more skill and confidence will form as a result. When forming the solution to a problem, write out each of the steps. Don't simply do head work. By doing as many steps on paper as possible, much clarification and therefore confidence will be formed. Do this with as many homework problems as possible, before checking the answers. By checking the answer after each problem, a reinforcement will exist, that will not be on the exam. Study situations should be as exam-like as possible, to prime the test-taker's system for the experience. By waiting to check the answers at the end, a psychological advantage will be formed, to decrease the stress factor.

Another fantastic reason for not cramming is the avoidance of confusion in concepts, especially when it comes to mathematics. 8-10 hours of study will become one hundred percent more effective if it is spread out over a week or at least several days, instead of doing it all in one sitting. Recognize that the human brain requires time in order to assimilate new material, so frequent breaks and a span of study time over several days will be much more beneficial.

Additionally, don't study right up until the point of the exam. Studying should stop a minimum of one hour before the exam begins. This allows the brain to rest and put things in their proper order. This will also provide the time to become as relaxed as possible when going into the examination room. The test-taker will also have time to eat well and eat sensibly. Know that the brain needs food as much as the rest of the body. With enough food and enough sleep, as well as a relaxed attitude, the body and the mind are primed for success.

Avoid any anxious classmates who are talking about the exam. These students only spread anxiety, and are not worth sharing the anxious sentimentalities.

Before the test also involves creating a positive attitude, so mental preparation should also be a point of concentration. There are many keys to creating a positive attitude. Should fears become rushing in, make a visualization of taking the exam, doing well, and seeing an A written on the paper. Write out a list of affirmations that will bring a feeling of confidence, such as "I am doing well in my English class," "I studied well and know my material," "I enjoy this class." Even if the affirmations aren't believed at first, it sends a positive message to the subconscious which will result in an alteration of the overall belief system, which is the system that creates reality.

If a sensation of panic begins, work with the fear and imagine the very worst! Work through the entire scenario of not passing the test, failing the entire course, and dropping out of school, followed by not getting a job, and pushing a shopping cart through the dark alley where you'll live. This will place things into perspective! Then, practice deep breathing and create a visualization of the opposite situation - achieving an "A" on the exam, passing the entire course, receiving the degree at a graduation ceremony.

On the day of the test, there are many things to be done to ensure the best results, as well as the most calm outlook. The following stages are suggested in order to maximize test-taking potential:

Begin the examination day with a moderate breakfast, and avoid any coffee or beverages with caffeine if the test taker is prone to jitters. Even people who are used to managing caffeine can feel jittery or light-headed when it is taken on a test day.
Attempt to do something that is relaxing before the examination begins. As last minute cramming clouds the mastering of overall concepts, it is better to use this time to create a calming outlook.
Be certain to arrive at the test location well in advance, in order to provide time to select a location that is away from doors, windows and other distractions, as well as giving enough time to relax before the test begins.
Keep away from anxiety generating classmates who will upset the sensation of stability and relaxation that is being attempted before the exam.
Should the waiting period before the exam begins cause anxiety, create a self-distraction by reading a light magazine or something else that is relaxing and simple.

During the exam itself, read the entire exam from beginning to end, and find out how much time should be allotted to each individual problem. Once writing the exam, should more time be taken for a problem, it should be abandoned, in order to begin another problem. If there is time at the end, the unfinished problem can always be returned to and completed.

Read the instructions very carefully - twice - so that unpleasant surprises won't follow during or after the exam has ended.

When writing the exam, pretend that the situation is actually simply the completion of homework within a library, or at home. This will assist in forming a relaxed atmosphere, and will allow the brain extra focus for the complex thinking function.

Begin the exam with all of the questions with which the most confidence is felt. This will build the confidence level regarding the entire exam and will begin a quality momentum. This will also create encouragement for trying the problems where uncertainty resides.

Going with the "gut instinct" is always the way to go when solving a problem. Second guessing should be avoided at all costs. Have confidence in the ability to do well.

For essay questions, create an outline in advance that will keep the mind organized and make certain that all of the points are remembered. For multiple choice, read every answer, even if the correct one has been spotted - a better one may exist.

Continue at a pace that is reasonable and not rushed, in order to be able to work carefully. Provide enough time to go over the answers at the end, to check for small errors that can be corrected.

Should a feeling of panic begin, breathe deeply, and think of the feeling of the body releasing sand through its pores. Visualize a calm, peaceful place, and include all of the sights, sounds and sensations of this image. Continue the deep breathing, and take a few minutes to continue this with closed eyes. When all is well again, return to the test.

If a "blanking" occurs for a certain question, skip it and move on to the next question. There will be time to return to the other question later. Get everything done that can be done, first, to guarantee all the grades that can be compiled, and to build all of the confidence possible. Then return to the weaker questions to build the marks from there.

Remember, one's own reality can be created, so as long as the belief is there, success will follow. And remember: anxiety can happen later, right now, there's an exam to be written!

After the examination is complete, whether there is a feeling for a good grade or a bad grade, don't dwell on the exam, and be certain to follow through on the reward that was promised...and enjoy it! Don't dwell on any mistakes that have been made, as there is nothing that can be done at this point anyway.

Additionally, don't begin to study for the next test right away. Do something relaxing for a while, and let the mind relax and prepare itself to begin absorbing information again.

From the results of the exam - both the grade and the entire experience, be certain to learn from what has gone on. Perfect studying habits and work some more on confidence in order to make the next examination experience even better than the last one.

Learn to avoid places where openings occurred for laziness, procrastination and day dreaming.

Use the time between this exam and the next one to better learn to relax, even learning to relax on cue, so that any anxiety can be controlled during the next exam. Learn how to relax the body. Slouch in your chair if that helps. Tighten and then relax all of the different muscle groups, one group at a time, beginning with the feet and then working all the way up to the neck and face. This will ultimately relax the muscles more than they were to begin with. Learn how to breathe deeply and comfortably, and focus on this breathing going in and out as a relaxing thought. With every exhale, repeat the word "relax."

As common as test anxiety is, it is very possible to overcome it. Make yourself one of the test-takers who overcome this frustrating hindrance.

Special Report: Additional Bonus Material

Due to our efforts to try to keep this book to a manageable length, we've created a link that will give you access to all of your additional bonus material.

Please visit http://www.mometrix.com/bonus948/pecivil to access the information.